D0897604

Chasing Lemurs

My Journey into the Heart of Madagascar

Keriann McGoogan

Prometheus Books

Guilford, Connecticut

 Prometheus Books

An imprint of The Rowman & Littlefield Publishing Group, Inc.
4501 Forbes Blvd., Ste. 200
Lanham, MD 20706
www.rowman.com

Distributed by NATIONAL BOOK NETWORK

British Library Cataloguing in Publication Information available

Library of Congress Control Number: 2019955295
ISBN 9781633886209 (cloth)
ISBN 9781633886216 (ebook)

∞™ The paper used in this publication meets the minimum requirements of American National Standard for Information Sciences—Permanence of Paper for Printed Library Materials, ANSI/NISO Z39.48-1992.

For Travis: You are my constant inspiration.
And for my Mom and Dad—my biggest champions.

Contents

PART IV

Prologue: Where There Is No Doctor

June 23, 2006

The sand gave a little under my hiking boots as I clambered up onto some large, exposed rocks. I gripped the handle of the small, black, waterproof Pelican briefcase that held the satellite phone—our only lifeline out here in the remote forests of northwestern Madagascar. From up high, I could see the twists and turns of the Mahavavy River as it made its way through the lush green forest—a spectacular view. But now wasn't the time to drink in the beauty of my surroundings.

I sat down, opened the case, and pulled out the phone. I watched the screen, waiting for the satellites to connect, and took a few deep breaths. Three hours before, in pitch darkness, I had opened my eyes. "Keriann." A voice was calling me. Louder: "Keriann." I didn't know where I was. I wasn't in my own bed. The sounds, the air—this didn't feel like Toronto. Where was I? As I stared into the dark, I began to see shapes. Directly above, I could see crooked tent poles propping up the fly of my two-man tent. That was it. I shifted my weight and felt the hard ground beneath my blow-up Therm-a-Rest mattress. As I went to reach for my glasses (I'm blind without them), I found that I was still securely zipped into my mummy-sack sleeping bag. Inside the bag, I fumbled for the zipper. Then that voice again, more urgent than before. "Keriann?"

All at once, I was back. The voice belonged to Fidèle, my happy-go-lucky Malagasy cook, who had accompanied me and the others into the remote forest of northwestern Madagascar. We had pitched our tents in a secluded patch of trees near the shore of the Mahavavy River. As a primatologist in training, I had come here to study lemurs in their natural habitat for my PhD. And it had been quite an adventure so far. Just getting here—to the majestically named Kasijy Special Reserve—had cost us a ten-day hike and a great deal of angst. But we had made it. And just yesterday, I remembered, smiling at the thought, we had spotted several groups of lemurs bouncing through the trees like hyped-up versions of Spider-Man. It had been a slog, but my field research season had at last *really* begun.

But something was wrong. What was Fidèle doing at my tent? I broke free of my sleeping bag, popped on my glasses, and felt around in the dark for my headlamp. For Fidèle to approach my tent so late at night was highly unusual. Malagasy people, I had learned, are extremely polite, and there is no way Fidèle would wake up the *vazaha* ("foreign") girl in the middle of the night unless he faced an emergency.

"Un moment," I said, and unzipped the door of my tent. As I emerged into the night, the bright light of Fidèle's headlamp blinded me momentarily. I blocked the glare with my hand and Fidèle, realizing his error, quickly dimmed his light. He didn't speak French or English—only Malagasy—and so he gestured for me to follow him toward Andry's tent. Andry, my Malagasy field assistant, was a student from the University of Antananarivo and had already worked a few seasons in the Madagascar wilds. He was experienced in the field and a pro at negotiating the local politics. Not only that, he was indispensable as a translator. None of my Malagasy field guides spoke English or French.

My supervisor, Shawn, had journeyed here with us, but he had headed home some time ago, leaving me in charge, with Andry as my assistant. Whatever was happening now with Andry couldn't be good. My heart sank when I arrived at his tent. He lay flat on his back, shivering uncontrollably. I knelt down next to the open tent-door and spoke softly. "Andry, what's wrong? Are you okay?"

"Ahh . . . non," Andry said, reverting to French. His teeth were chattering, and his breathing was shallow. "I am sick, Keriann. Very sick."

"Hang on," I said. My mind racing, I dashed back to my tent to grab my medical bag. I didn't know what I could do for Andry. I was no doctor. I grabbed my kit and a book that Shawn had left with me, appropriately titled *Where There Is No Doctor*.[1] This book, Shawn had said, is a guide to people's health in remote locations. Topics range from diarrhea to ringworm to malaria. It explains how to prevent, recognize, and treat common ailments when you are in places where, as the title indicates, there is no doctor. I had hoped I wouldn't need to use this book, but here we were, only a few weeks into my research season.

Back at Andry's tent, and with my headlamp on, I asked him about his symptoms as I paged slowly through the book. Fever. Chills. Aches. Unfortunately, most of the symptoms were pretty general and could be a sign of almost anything. Sweats. Headache. Nausea. I had a hunch but didn't want to think it.

This Madagascar journey wasn't my first foray into fieldwork. While completing my master's degree at the University of Calgary, I had studied the impact of a devastating Category 4 hurricane on the population density of black howler monkeys in Belize, Central America. I had lived and worked in Belize for six months. And yes, I had dealt before with getting sick in the field.

In fact, we grad students had made a game of trying to diagnose ourselves using our Lonely Planet health guide. Back then, when we were only moderately ill, we had almost laughed as we read through the symptoms and came to the consistent conclusion that we had something terrible. Dengue fever for sure. Or just as bad: malaria. Hilarious, we thought. But now, in Madagascar, with Andry shivering in his tent, no doctor within reach, no one around that spoke either of my languages, and no easy way back to civilization, my memory of that game had lost its charm.

"Keriann," Andry said. I looked up from the pages of the book. "I think it might be malaria."

Now, on the ridge, as I waited for the satellites to link up, I gazed out over the forest and the winding Mahavavy River and told myself that everything would be okay. At last the final satellite appeared on the screen and I dialed a number in far-distant Canada. After a short pause, I heard the ring tone.

Bzzzzzzz. Bzzzzzzz. "Hello?"

My heart leapt, and I jumped to my feet when I heard the friendly female voice on the other end of the line. Shawn's wife. She sounded strangely close, as if I were talking to her from my home in Toronto.

"Hi. Is this Christine?" I asked, trying to mask my sense of urgency.

"Yes."

"Christine, this is Keriann. Shawn's PhD student? We met a few weeks ago when you dropped off the extra batteries for Madagascar. I'm just at our field site on the satellite phone. Is–is Shawn home yet? I know he was supposed to fly out yesterday, and I—"

"Keriann," Christine cut me off in a kind voice, "Shawn is still in transit. His second flight got delayed. He won't arrive home until day after tomorrow. Is everything okay?"

"Oh, I didn't realize. Of course, he's not home yet. Umm, well." How much to tell her? "It's about our Malagasy student assistant. It's just—well, he's got a very high fever and is shivering uncontrollably and I'm—"

"Uncontrollably? What do you mean?" Christine asked. I could hear concern mounting in her voice.

"I'm pretty sure he's got malaria," I continued calmly. "He took some medicine, possibly for malaria, but I'm not sure what it is—there's no label. I have been reading our medical book about cerebral malaria and—"

"Who else is there with you?"

"Well, that's the problem. Besides me, Andry is the only other person here who speaks English or French. The rest of the team are Malagasy from the remote communities. The other student and the gendarmes left with Shawn a few days ago."

"Shawn said it was a ten-day hike just to get in there," Christine said. "And didn't he say something about a warning? Something about bandits in the area?"

I squatted down on the rocks and rested my head in my hand.

"Keriann? Are you still there?"

I looked out over the towering trees that lined the banks of the Mahavavy. There were lemurs in that forest. I had spotted them a few days ago while washing my laundry in the river. Dozens of lemurs, scores, probably hundreds. And we had only just begun my field research season. "Not to worry," I told Christine. "I had hoped to double-check with Shawn. But I know what I have to do."

Basically, I told a bare-faced lie.

Introduction

The island nation of Madagascar, located off the east coast of Africa, has been isolated from the mainland for at least 130 million years.[1] Its biodiversity is unique. Madagascar is home to insects, reptiles, birds, and lemurs that are found nowhere else in the world. I have had the privilege, while doing research for my PhD in biological anthropology, of living and working in Madagascar for nineteen months.

During my three sojourns, I worked twelve-hour days in the forest chasing groups of lemurs. I learned how to speak Malagasy. I hiked hundreds of kilometers through remote forest fragments. I visited dry forest, cloud forest, and beaches. I worked at a field site where there was no access to water. Twice I got malaria. I saw the extent of Madagascar's forest loss firsthand and lamented the extinction of the island's giant lemurs. I saw Malagasy people—people like you and me, but less lucky in the land of their birth—living in abject poverty and dying from preventable diseases.

In 2011, I received my PhD from the University of Toronto. Since then, I have wanted to write something more personal about Madagascar. When, finally, I sat down to do so, I found myself reliving my first trip into the wilds of the country. I was twenty-five years old. I wasn't soul searching or seeking to "find myself"—I went to study lemurs in their natural habitat, and to set up a permanent field site in the remote northwest—a site to which I could later return to do further research. Although my PhD supervisor, Shawn Lehman, would be with me for the first few weeks of the trip, I packed my duffle bags knowing I would be going it alone for the majority of the time, and I would be the only female on the trip.

Despite careful planning and preparation, the trip spiraled out of control. Food poisoning, impassible backcountry roads, grueling hikes, challenging local politics, malaria—these would turn a simple reconnaissance outing into an epic field test. I did not anticipate that we would encounter a roaming band of thieves, or that I would wind up among men who spoke nothing but Malagasy and leading them on a forced rescue mission. I didn't set out to find myself. I wasn't soul searching. Yet somehow, against the staggering backdrop of the Madagascar wilds, I was forever altered.

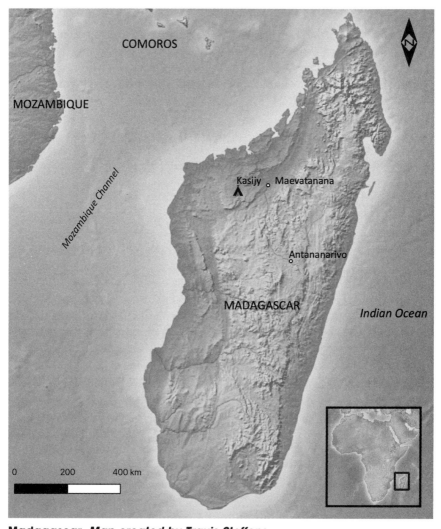

Madagascar. *Map created by Travis Steffens.*

Part I

Toronto, Canada. *Map created by Travis Steffens.*

CHAPTER 1

Jane Goodall Is Not Alone

April 2006

Over Shawn's shoulder I could see the Google Map satellite image on the computer screen.

"There it is," he said, using his computer's mouse to point to the familiar location marker on the screen—a red pushpin nestled among a sea of green. "Kasijy Special Reserve." He looked over his shoulder at me. "It's still incredible to me that there's only been two other teams of researchers who've worked there—and they were in and out in a few weeks as part of a rapid assessment of the biodiversity."

My PhD supervisor, Dr. Shawn Lehman, had invited me for a "debriefing" meeting a few weeks before we departed, separately, for my three-month preliminary field research trip to Madagascar. We were meeting to discuss trip logistics, project setup, and any other questions that I had. We were searching for a new field site in Madagascar—a site where I could return to conduct my PhD research on lemurs, research that would span more than a year.

Shawn had previously worked at a site in southeast Madagascar, but tensions with the local people had forced him to abandon that field site. And so, for the past year, Shawn had been poring over satellite images and published literature looking for a new site and had finally settled on Kasijy. We would go there together, set up the research camp, and collect some data while we were at it. As I squinted at the satellite image of what was to become my new field site, I found it hard to believe that just nine months before, I had sat in this same office for the first time.

"Call me Shawn," he had said as we shook hands. I was twenty-three years old and had come to meet Shawn before he officially became my PhD supervisor at the University of Toronto. I was nervous. I had admired Shawn's work on primate conservation biogeography when I was completing my master's at the University of Calgary, and I was excited to become a part of his new research program focusing on the impact of forest edges on lemur species in Madagascar—*edge effects* to those in the know. My master's supervisor had only positive things to say about Shawn and his work, but even so, I couldn't seem to calm the butterflies, and Shawn's imposing stature did nothing to ease my nerves.

Shawn had been a varsity football player at university, and he looked it. At five feet seven, I am not a short woman, but Shawn is about six-five, and he towered over me when he stood to shake my hand. I had to stifle a gasp. This was my new supervisor? Did I accidentally sign up for a degree in kinesiology? After about twenty minutes of small talk about our shared interests and background—we had both lived in Calgary and had a passion for primate conservation—I grew calmer and more comfortable. Shawn was an affable guy with a great sense of humor, and he seemed as excited as I was about my joining his team. In fact, he proudly noted that I would be his first PhD student.

Now it was April, and I had finished the first year of my program. It had been a hectic eight months. I had moved from Calgary to Toronto—away from my long-term partner, now fiancé, Travis, who was still living in Calgary. I had taken a full graduate-student course load, written grant proposals, worked as a teaching assistant, and passed a required language exam. Those life events and academic rites of passage now behind me, I was excited to get to Madagascar and start on the next chapter.

I had wanted to go to Madagascar ever since my first primatology course as an undergraduate at the University of Calgary, Introduction to Physical Anthropology, which I had taken as an elective during the first year of my English degree. I would like to note that I had formed this desire before Disney's *Madagascar* animated film and the comical King Julien ever existed. In that course, we had learned all about lemurs, a type of primate found only in Madagascar. Our instructor, Brian Keating, was the head of Conservation Outreach at the Calgary Zoo and was locally famous in Calgary—I recognized him from his occasional appearances on the Discovery Channel and radio interviews.

Brian taught us that Madagascar is situated in the Indian Ocean off the southeast coast of Africa. It is the fourth largest island in the world—after Greenland, New Guinea, and Borneo—and roughly the size of France.[1] And it is unique—a world unlike any other. Madagascar has been separated from other landmasses for eighty-eight million years, and from mainland Africa for at least one hundred thirty million.[2] As a result, evolution has occurred there in a microcosm. Madagascar has more endemic species than any other country. More than 480 genera, or groups of species, and twenty-six families are found only here and nowhere else on earth.[3]

Madagascar combines these record levels of endemism with astonishing diversity. For example, the country is home to over eleven thousand native species of vascular plants.[4] Of these, 83 percent are endemic.[5] Visit Madagascar and you'll also find an impressive 457 species of reptiles, with 96 percent endemic.[6] There are 503 species of bird, and 60 percent of those are found nowhere but there.[7] And I have not even mentioned Madagascar's main claim to fame: its lemur population. The country is home to 111 species and subspecies of lemurs, 100 percent of which are found only there.[8]

In Brian's course, we learned that each of the different species of lemur is singular and amazing. Take the Indri, for example—the largest of all living lemurs at six kilograms,[9] whose haunting call is matched only by its hilarious appearance (think of a two-year-old in a panda suit). Their loud whooping calls or songs are otherworldly, high-pitched but guttural, and can be heard up to four kilometers away. The Indri uses these calls for group spacing—as if to say, "Hey, other groups. We're over here. We're going to be eating these figs for a while. Steer clear."

Then there is the aye-aye. These elusive, nocturnal lemurs have long, bony middle fingers that they use to tap on hollow trees or bamboo stalks while hunting for insects and grubs, just as woodpeckers do with their beaks. Aye-ayes then use their mobile ears to listen for prey. If they find something, they drill a hole in the tree using their ever-growing incisor teeth—similar to what you see in rodents—and scoop the insect out with their long middle digit. I watched in that two-hundred-seat lecture hall, with the wide-eyed wonder that only a first-year undergraduate can muster, as the alien-looking aye-aye on the screen nabbed its prey.

Brian was larger than life, and he opened my eyes to creatures and concepts I had never known existed. Suddenly, I felt . . . it. A tangible shift. A shift in the lecture hall, and a shift in myself. A light went on, and it felt electric. Brian delivered every word that he spoke with authentic enthusiasm, an enthusiasm so intense that it was infectious. We students could feel his passion. I didn't stand a chance.

The class inspired me to pay a few visits to the public library, where I checked out various nature documentaries. I was obsessed. I learned that lemurs aren't all that Madagascar has to offer. There are giraffe-necked weevils, for example—tiny insects with bright red wing covers and long black necks, like giraffes, which make them look like an alien species. Males have necks that are two to three times longer than those of females, and they use these to fight other males for mating access—sexual selection at its most remarkable. Then there are leaf-tailed geckos, known for mimicking dead leaves and blending in seamlessly with their surroundings. But that's not all. Near the beaches in Ile Sainte Marie, a tiny island off the east coast of Madagascar, there are humpback whales to be found, dancing through the waters of the Indian Ocean.

Lemurs and insects and birds, oh my!

"But here's the clincher," Keating said one afternoon, near the end of the semester. "Nearly all of Madagascar's lemurs are threatened with extinction. Think about that for a moment." My mind raced. He was right, of course. My research at the library had confirmed it. Madagascar is a country at risk—a country in peril. In their guide *Lemurs of Madagascar*, Conservation International points out that because of the country's combined diversity and endemicity, "a hectare of forest lost in Madagascar has a greater negative impact on global biodiversity than a hectare lost almost anywhere else."[10] And, unfortunately, Madagascar has experienced extremely high levels of habitat loss. Researchers estimate that about 50 to 60 percent of Madagascar's forest disappeared during the first five hundred years of human occupation.[11] Between 1953 and 2014, the island lost an additional 44 percent of its natural forest cover.[12] This loss can be attributed to human actions including farming, mining, and logging. What remains of the vegetation is highly fragmented.

As might be expected with these levels of habitat loss, Madagascar has already seen extinctions of some incredible biodiversity. The first explorers to visit Madagascar would have seen far more species than exist today. Extinctions include eight species of elephant bird, three species of pygmy hippopotamus, two species of giant tortoise, and at least seventeen additional species of giant lemurs—some the size of gorillas.[13] When I first learned about the giant lemurs that once lived in Madagascar, I felt both fascinated and devastated—fascinated that such creatures had once existed and devastated to realize that they had survived well into the Holocene (the geological epoch we are in now), with the last-occurrence dates estimated between just 2,400 and 500 years before present.[14] I had missed them by a few hundred years—a minuscule amount of time in evolutionary terms.

But, as Brian had shown us, the issue was more complicated than I could ever imagine. With a population of twenty-five and a half million,[15] Madagascar is one of the poorest countries in the world. Of the sixteen million people who live in rural areas, some 85 percent live below the poverty line, often relying on tiny plots of farmland for survival, and subsisting on less than two dollars a day.[16] People cut down trees simply to feed their families. UNICEF estimates that only 11 percent of the total population has adequate sanitation, and only 29 percent of rural families have access to clean drinking water.[17] Prominent lemur researcher Alison Jolly said it best when she wrote that Madagascar's plight is "a tragedy without villains."[18]

In our last class that very first year of my undergraduate degree—a time when I was probably at my most impressionable—Brian made his final plea. On the screen, he projected an image of an Indri.

"It would be a sad world without these amazing creatures. You are the next generation. It's in your hands to make a difference."

We broke out into a round of applause. Brian Keating's course prompted me to change my major from English to anthropology and ultimately led me to that moment in Shawn's office, planning our trip to a faraway part of the world.

Shawn turned back to the map on the screen. "Kasijy is extremely remote. Look here." Shawn zoomed in. "That's the nearest large village—Kandreho." He pronounced the village name as "kan-dree-hoo," as the letter "o" in Malagasy is always elongated. "It's about thirty-two kilometers from Kasijy as the crow flies. We'll stop there, and then I think we'll be able to drive—slowly—most of the way. We'll hit a point, though, where we will have to get out and walk—probably a one- or two-day hike. It's going to be rocky, sandy terrain—a tough hike—and we'll be camping in the bush when we arrive. We'll be working with the local villagers, who speak only Malagasy. In fact, this area is so remote that we might be the only Westerners they've ever seen. Should be interesting." He turned to me and raised his eyebrows. "But you're up for it, right?"

I hesitated. *Was I up for it?* I wanted to think so. Researching nonhuman primates in their natural habitat wasn't new to me. In 2003, I finished my undergraduate degree—a bachelor of science in primatology. (A quick pause to say what we are all thinking: *Who knew a degree in primates was a thing?* But, yes, it is so.)

After being inspired by Brian Keating, I scoured the course calendar and was delighted to discover not just one but many courses that focused on nonhuman primates. The University of Calgary was unique in Canada at the time because it offered a specialized program in primatology. While many Canadian anthropology departments were lucky to have just one faculty primatologist, the University of Calgary had four. One of these academics was Dr. Mary Pavelka. Mary's lecturing style is engaging, inspiring, and informative—hitting all three of those buckets is no small feat. From Mary, I took one of my most memorable courses as an undergraduate student—the History of Primatology. In that course, I learned more about the discipline of primatology, and the surprising number of researchers who had come before me. One of the most famous primatologists that we covered was, of course, the inspirational Dr. Jane Goodall. Goodall, I learned, had set out into the wilds of Gombe Stream National Park in Tanzania way back in 1960 to study chimpanzees. She was only twenty-six years old.[19] She had been sent to Gombe by Louis Leakey, a famous paleontologist who believed that studying living nonhuman primates would provide insights into our early human ancestors. Goodall spent years with the chimpanzees, observing and writing pages and pages of notes in her field journal. She discovered so much that we didn't know about primates in the wild and about chimpanzees in particular—chimps were omnivorous, not vegetarian; they hunted small animals; they used tools; and they even engaged in warfare with one another.[20] Goodall's work taught us the value of research on primates in the wild and her Gombe chimpanzee field site is the world's longest-running continuous wildlife research project.

At the end of the third year of my undergraduate degree, I found out that Mary was running a three-week field school in Belize, and I knew I had to be there. Here was a chance for my first foray into primatological fieldwork. A chance to access my inner Jane Goodall. The field school was held near the aptly named town of Monkey River, home to black howler monkeys. Howler monkeys, as the name implies, are famous for being among the loudest of any terrestrial land mammal.[21] The field school was a carefully planned mixture of time spent studying howler monkey behavior, learning about the forest, implementing primatological research methods, and indulging in some fun activities and field trips, including a reef snorkeling trip. It was a chance to get your feet wet, and to get a taste for real fieldwork.

I graduated with my bachelor of science and was accepted into the master's program in anthropology at the University of Calgary to work with Mary. Rather than find a traditional summer job between semesters, I spent another four months in the Belizean forests as a research assistant for two graduate students. My master's project focused on the impact of a devastating Category 4 hurricane on the population density of black howler monkeys in Monkey River. I lived and worked in Belize for a six-month stretch. I spent weeks following monkeys around, and even earned myself a nickname among the local Belizean people. I am and forever will be . . . *monkey girl*. I battled insects and reptiles and kayaked rivers surrounded by tropical forests chock-full of biodiversity. I also got to know some amazing local Belizeans who welcomed me into their homes and their lives.

Many graduate students at the University of Calgary who were based at more "comfortable" tropical sites would marvel at those of us working in Belize, where the field was a challenging beast. I dealt with otherworldly numbers of mosquitoes and biting flies, the threat of venomous snakes and scorpions, the physical exertion involved in cutting trails through thick swaths of forest, torrential rains, and seasonally flooded forests. Some days the rain would cause the river to flood and we would spend six to eight hours wading through the murky hip-deep water in the forest while we tracked the monkeys. Trench foot was not uncommon. All this experience—the good and the bad—had fostered a love of the field in me. Shawn's question echoed in my brain: *Are you up for it?*

I knew there would be key differences between my experience in Belize and what I would face in Madagascar. Belize is a former British colony, for starters.[22] Almost all the local people speak English, except for some Spanish speakers who had arrived from nearby Guatemala. Even the local Belizean Creole dialect is rooted in English, and by the end of my six-month field season I could understand about 70 percent of what the locals were saying when they spoke Creole. Traveling in an English-speaking country is relatively easy. Lost? Just ask directions. Hungry? Follow the signs to the nearest restaurant or grocery store. Sick? Visit the doctor and explain what is ailing you.

Madagascar, on the other hand, is a former French colony.[23] In 2006, English was added as one of three official languages: Malagasy, French, and English (note that the status of English as an official language was withdrawn in 2010),[24] but the majority of people speak only French and Malagasy. And, outside of the capital and larger cities, most speak only Malagasy. Along with my PhD courses that first year, I had taken a series of classes at the Alliance Française on weekends to brush up on my rusty high school French. I had passed my PhD language exam in French (a requirement for the program) with flying colors. "I don't think I would score 96 percent on a test about my life," Shawn joked when he received my grade notification. Even so, I was more conversational than fluent, and I did not speak a lick of Malagasy.

Second, the field site in Belize had been moderately comfortable. I lived on the top floor of a double-decker cabin next to the Monkey River. Mind you, this was not a place with all the modern amenities. It had wooden floors and concrete walls, and we discovered the hard way that it wasn't sealed well enough to keep out the creepy crawlies—and there were plenty of those. But it did have beds with legitimate mattresses, mosquito netting, a flush toilet, a generator for electricity, and a gas stove for cooking. The graduate students who came in and out of the site would also leave behind various "luxuries," such as DVDs, a battery-operated CD player, and a complete collection of *Harry Potter* books. The nearby beach town of Placencia was familiar from my stay during field school, and once a month we would hitch a ride with one of the local villagers in a speedboat. We would spend one or two nights in a hotel, stock up on groceries, watch television, and gorge ourselves on Snickers bars and Pringles.

In Madagascar, I would be camping in the bush. When we arrived at Kasijy, we would look for a spot close enough to the river but also sheltered by the forest so that we would have shade. We would dig our own latrines and would need to bring water filtration systems with us. Unlike when I was in Belize, any cooking we would do would be over an open fire. There would be no supply runs after we arrived at the site. We would have one chance to bring in all of our food and supplies, because to get to the nearest village would involve a multiday hike. There would be no air-conditioned hotels in the vicinity.

And there was one other important difference. In Belize, my partner, Travis, had lived with me as my field assistant. Travis and I first met on the Belize field school. He was also working to become a primate researcher, and he had found his calling in fieldwork. Six feet tall, with a slim, athletic build and sandy brown hair, Travis is big lover of the outdoors and adventure. He is at his best when the challenges of the field are at their worst. In Belize, he could muster determination and optimism when others just wanted to quit. He could turn things around.

I remember one day when the going got tough. To contribute to the long-term data we were collecting at the Monkey River field site, we had been tasked with finding and recording the location of each of four monkey groups every day. We had struck out to find our monkey groups at five a.m. Often, the monkeys would move around in the early part of the day, making them easier to find. By ten a.m., though, we had not found any groups whatsoever. It was getting hot, and we were tired from walking through the forest in circles, probably passing the monkeys without realizing it. I had had enough. I sat down on a log near our cabin and cried with frustration. Travis sat down next to me on that log in Belize and gave me a pep talk that was sweet, funny, and motivating all at once.

"Look at where we are," he said, gesturing to the lush green forest around us. "A lot of people would kill for this opportunity. We couldn't find the monkeys today—so what? We tried our best, and tomorrow we'll go out again. We'll find 'em." He put his arm around me and gave me a squeeze. We got back up, and later that day we did locate monkeys. Having Travis with me in the field made the day-to-day challenges so much easier to handle.

But Travis wasn't coming to Madagascar with me. He was a year behind me and was about to commence the second year of his master's degree at the University of Calgary. Before I applied to the PhD program in Toronto, Travis and I had sat in the front seat of his mother's old red minivan and discussed what was next for us.

"I've always wanted to study lemurs," I told him. "Ever since Brian's class. There's a researcher at the University of Toronto. Shawn Lehman. His research site is in Madagascar, and he is interested in questions about biogeography and conservation. My parents are living out that way too—I could stay with them for free and save some money. It would be ideal."

"You do realize that I can't move with you because I have to finish my master's here," he said, gently. "Plus, I have to go to Belize for six months for my fieldwork. You wouldn't be able to visit me easily there."

We looked at each other, meaningfully.

"That's true." I paused, digesting his words. What he was saying felt a little too real. But I couldn't shake the feeling that I needed to do this. It was time. I needed to branch out on my own. Plus, it was *Madagascar*. We agreed that I needed to go for it.

As my departure to Madagascar now loomed, Travis was back in Belize, gathering data for his thesis. Typically, he had opted to do a difficult project and was traveling the entire country by motorcycle to look for monkeys. He wanted to determine the habitat features that were associated with the distribution of black howlers in Belize. We had spent the entire year apart after I'd moved from Calgary to Toronto. It had been a long year, and we had managed to visit each other only twice. He had come to Toronto once in the fall and I had returned the favor by meeting up with him in Belize in the spring, just before my Madagascar trip.

I was excited to see Travis before I left, and I figured that while I was in Belize, I would help him out with his research by tagging along on some of his monkey surveys. We spent a week tooling around on Travis's motorcycle, stopping in at various forest patches to look for black howlers. This resulted in a few misadventures. On one occasion, we got a flat tire on a dirt road near an orange grove in the middle of nowhere. We had no spare or patch kit, but luckily spotted a few Spanish workers in the grove. Travis politely approached them and tried to explain in his broken Spanish what had happened.

"Tire . . . PSHHHHH!"

They laughed at his noble attempt to communicate and directed us to another worker who did speak English. The force must have been with us because in fifteen minutes, someone was coming to collect the orange harvest from this location. So, we waited. Sure enough, fifteen minutes later a big tractor arrived, hauling a trailer to collect the oranges. The driver said he would fill up the trailer and then drive us to someone nearby who could help us patch the tire. A small miracle! As we sat perched atop a trailer full of Valencia oranges, we looked at each other and burst out laughing. To this day, I think about that every time I drink orange juice. You just never know who might have sat on those oranges before they became juice.

A few mornings later, we set out to camp for three days while searching for monkeys. The night before, we had packed up all our camping gear, including a tent, and loaded up the motorcycle to drive to the trailhead. I was balanced on the back of the bike with a giant backpack filled with camping gear. I had to hang on to Travis for dear life, and I was not comfortable. We had to pull over a few times so that I could take off the backpack and rest my shoulders.

Upon arriving at what Travis said was our destination, we started hiking through a forested area. It was a pine forest, which I thought a bit strange—howler monkeys don't typically live in pine forest. I looked around and said, "Are you sure there are monkeys here?"

Travis assured me that he'd heard a rumor about monkeys living in this vicinity, and so we marched on. It was scorching hot and humid, and the hike involved a lot of uphill slogging. We had each brought only one liter of water because Travis had said that we would be able to fill up our bottles at our destination. After about an hour of slogging, I had finished most of my water. I was getting frustrated, and Travis was clearly not putting in a lot of effort to look for monkeys.

"You're not even looking up," I said, exasperated.

Abruptly, Travis turned and said that we needed to go back to the parking lot to check a different trail. So back we went. I was tired, hot, and aghast that

Travis was forcing me to carry on. I thought he loved me? We reached the second trailhead and as we hiked down a large hill, some small pools and waterfalls came into view. Next to the falls, I could see a picnic table set with champagne glasses and a cooler full of food. Travis marched directly to the table and gestured for me to follow.

I stopped in my tracks.

"Trav, come back," I called. "That is somebody else's picnic! We shouldn't even be here."

He took my hand and led me to the table.

"This whole day has been a lie," he said.

He had brought me here—to a space on the property of a gorgeous resort hotel—to ask me to marry him. Travis got down on one knee and presented me with his grandmother's wedding ring. What could I do? I said yes.

Sitting in Shawn's Toronto office, as we prepared for our trip to Madagascar, that beautiful afternoon in Belize felt like a distant memory, even though it was only a few weeks before. I looked down at the diamond ring that now had a home on my left ring finger. *I'm engaged*, I thought and almost gasped. The whole situation felt truly unbelievable. Here I was, about to depart on my first big trip alone, with no Travis to back me up. Shawn's words again: *Are you up for it?*

I remembered the idea of the imposter syndrome, which I'd found in a book by the feminist psychotherapist Pauline Clance. She conceived this as having an "internal experience of phoniness" and "feeling like a fraud."[25] I couldn't recall the details, but Clance argued that this trait turns up in humans of both genders, but especially among high-achieving women. But as I have since learned, the fear of "being found out" is not limited to women. I can point to John Steinbeck, a Nobel Prize winner: "I am not a writer. I've been fooling myself and other people. . . . I hope this book is some good, but I have less and less hope of it."[26]

Up to now, I had always felt I was standing in Travis's shadow when it came to fieldwork. Travis shines in the field. He loves the outdoors and isn't afraid to feel uncomfortable. He seeks out adventure and challenge and is an expert in the backcountry, having camped out since boyhood. Me, on the other hand, if you had asked me even three years before whether I thought I could hack it on a backpacking trip into the remote forests of Madagascar . . . well, let's put it this way, my first real backpacking trip had taken place when I was twenty-two—the summer before I moved to Toronto. Travis and I had arranged to hike and camp the Iceline Trail in Yoho with a couple we had worked with in Belize. I was the least experienced camper of the group by far. Travis had been living and working at the Plain-of-Six-Glaciers teahouse in Lake Louise—a five-kilometer hike from the Chateau. For the entire summer, he worked five days on and two days off, and would literally jog up and down the mountain to and from the Chateau on his off days.

The other couple, Kyle and Aliah, were also outdoorsy and in peak physical condition. Kyle had been a competitive cross-country skier, and Aliah was no slouch when it came to physical exertion. While I had been working out at the university gym regularly, and had done my time in Belize, I was no match for this group, and I was thankful that Travis had taken the lead in preparing our gear and food for the trip.

We were going to hike the Whaleback circuit, which promised beautiful glacier views of the Yoho Valley and Twin Falls from both above and below. We would spend one night and two days hiking about twenty kilometers, with a 520-meter elevation gain. When we arrived at the trailhead, Travis, in his excitement, took off like a shot in the lead and the others quickly followed. I huffed and puffed, struggling with my heavy backpack, pulling up the rear. The hike was gorgeous, and the weather was perfect. We had hoped to camp at the halfway point of the loop, but all the sites in that location had been booked, so we reserved a spot farther on.

When we stopped for lunch, I swallowed my pride and confessed to the group: "I think we might need to slow the pace a little. I'm not sure that I can keep this up. Do you mind if we take a few more breaks?"

"Of course," Kyle said, kindly. Everyone understood, but I couldn't help feeling guilty about slowing down the group. This is one of the reasons that I have never particularly taken to team sports. I don't like being responsible for other people's successes or failures. I'll manage my own, thank you very much.

We forged ahead, tackling a lot of ups and downs, and by four p.m., I was nearing my limit. We had gone so fast (for me) in the beginning, and I didn't have much left to give. I found going down especially difficult with the heavy pack—something that I wasn't used to—and my frustration with being at the rear, struggling to keep up, was mounting. Eventually, I "bonked" and the tears came. I stopped in my tracks.

"I'm not sure I can do this," I called to Travis, who had been keeping pace for me up ahead.

He turned and saw that I had stopped.

"You can do this, Keriann," Travis said, encouragingly. "This is definitely a hard hike, but you got this. And when we get to camp, we'll have dinner. It'll be great. Here." He pulled an energy bar from his pocket. "Try eating something."

I took the bar and started eating. I took a few sips of water. Kyle and Aliah powered ahead, showing no signs of slowing down. I realized that I had to go on. I couldn't stop here.

"Is everything okay?" Kyle shouted to Travis when he saw we had stopped. He jogged back up toward us (show off, I remember joking to myself).

"Here. I'm still feeling good. Why don't I take your pack?"

Kyle kindly carried my backpack the rest of the way, and with Travis's encouragement, I did make it to our campsite. That trip was the first and only time that I had backpacked pre-Madagascar, but I would soon learn that the Iceline trail was good preparation for the physical and psychological challenges that at this point lay ahead.

Standing now in Shawn's office, I shook off the twinge of self-doubt. Fought back the imposter syndrome. It was time, I knew, to show Travis, and everybody else, how capable I had become. I thought about how Dr. Jane Goodall had been only twenty-six years old when she went to Gombe to study the chimpanzees. That was only two years older than I was now. Goodall had packed up her life and made for Gombe. She went by herself in 1960—a time when women were expected to get married and stay in the home.

"Yeah," I said, with more determination than ever. "I'm up for it."

"Grab a seat," Shawn offered after he'd finished marveling over his satellite map. "Let's talk logistics."

———————————

Email received from Travis Steffens, on April 26, 2006
Re: Hello from Belize

I am so jealous of your Madagascar trip! I can't wait for us to go there together for your full project. The lemur photos that Shawn forwarded look amazing. I was reading more about Von der Decken's sifaka online last night. It's an endangered species, but it hasn't been studied in the wild. It would be awesome if you were the first.

By the way, yesterday I drank some Valencia orange juice and I couldn't help but laugh out loud. The waitress at my usual restaurant must've thought I was crazy.

———————————

CHAPTER 2

Don't Forget the Batteries

May 2006

Shawn would fly to Madagascar first. He would arrive in Antananarivo (the capital city, nicknamed "Tana" after its previous French name, Tananarive) on May 12, and I would follow ten days later. Shawn planned to use the time before my arrival to work with our two Malagasy students from the University of Antananarivo, one of whom Shawn had worked with previously. Together, they would buy and pack most of the food and field supplies. I would arrive, and we would stay in Tana at the bed-and-breakfast he'd arranged so that I could acclimatize. After three days, we would hit the trail. Destination: lemurs.

"Almost no one has been to Kasijy, Keriann," Shawn said, excitedly. "It is practically untouched—the lemurs there are some of the least studied primate populations that exist in any protected area in northern Madagascar."

I nodded. I knew that a few research groups had visited Kasijy to conduct preliminary surveys of the lemurs.[1] These researchers had collected baseline data about which lemur species were there (we had eight different species awaiting), but we would be the first research team to spend more than a month in Kasijy. We would have the opportunity to answer new questions about the distribution and diversity of lemurs, lemur parasites, and vegetation in the reserve.

"We should be on the lookout for the crowned sifaka," Shawn continued.

My mind flashed to an image I had looked up on the internet. Crowned sifakas (pronounced "shee-*fak*-aaah") are characterized by beautiful, snowy-white bodies, with a hint of golden brown on their shoulders and a rich, dark chocolate–colored chest, back, and head. Like many lemur species, the crowned sifaka is listed as endangered by the International Union for Conservation of Nature.[2]

"Kasijy is the only protected area that contains crowned sifakas. Not only that, but the other research team that visited Kasijy spotted crowned sifakas in close proximity to another species of sifaka—Von der Decken's sifaka," Shawn said. "There could be some interesting data to uncover regarding the taxonomic classification of these two species."

I sat quietly, hiding my excitement. This trip could mean big things for my academic career.

"Here, take this," Shawn slid a neatly stapled stack of papers toward me. "I put this booklet together to share with my grad students who have never been to Madagascar. It outlines what to expect, from navigating the airport to getting supplies to starting up a lemur survey." He flipped it open and pointed to a table on the page. "See here," he said excitedly, "I've even provided a chart that breaks down how many kilograms of rice you'll need to buy, based on how much each person eats in a day." He grinned at me: "I hope you like rice. The Malagasy do love their rice."

Before you leave, make sure you consult a travel clinic to ensure that you have all your necessary immunizations and have been prescribed anti-malarial prophylaxis.

I reread the medical section of Shawn's "Unofficial Guide to Madagascar" as I sat in the waiting room of the university health clinic. Already I could see that this guide would be indispensable. I had made an appointment with the travel doctor to get all my prescriptions and shots before the trip. Ever since my field-work in Belize, I had dreaded going to travel doctors. They geared their advice to the average tourist, not to someone going to live in the destination country for several months. This is understandable. Most of their patients probably are headed to Caribbean all-inclusive resorts with swim-up bars. But I considered myself a seasoned traveler who had moved beyond basic travel advice.

"Keriann?" the nurse called from the front of the waiting room.

I followed her into the doctor's office and sat down in a chair next to the computer. In my mind, I ran through all the medications I would need for my Madagascar trip. Cipro for sure. My time in Belize had taught me that digestive ailments were more common in tropical locales. I would need to make certain that I got at least three courses of ciprofloxacin, an antibiotic used to treat bacterial infections, in case I got sick from the food or water. During my six-month stint in Belize for my master's research, I had been driven to Cipro more than once. We had mostly cooked for ourselves while we were there—a lot of canned corned beef and rice and beans—but on our monthly excursions to Placencia, we would sometimes get sick from the restaurant food.

I pulled my vaccination record out of the front pouch of my backpack and flipped through it. Hep A and B? Check. Tetanus? Yup. In fact, it looked as though I was up-to-date on most of my vaccines, except for typhoid. Oh yes, and I would need malaria meds too. In Belize, I had taken chloroquine, a relatively benign prophylaxis, and had not experienced a ton of side effects outside of some weird dreams from time to time. But the strains of malaria were different in Madagascar, so I would probably need a different medication.

The door swung open and a kind-looking, white-haired doctor came in smiling. He introduced himself as Dr. Turner and took the seat next to me at the computer. "So, you're planning a trip. Where are you headed?"

I told him that I was leaving to spend three months in Madagascar to do preliminary field research for my PhD. His eyes widened a little, "Madagascar? Well, you don't hear that every day." He reached up to a bookshelf above his computer and pulled down a gray binder. "Let's see." He flipped through the pages. "I'll start by reading through some travel hints that I borrowed from my friend who works at the Tropical Disease Clinic at the Toronto General Hospital. Don't get bit, don't get hit, and don't get lit," Dr. Turner recited.

I had to laugh. Was this guy for real? Dr. Turner looked up from his notes with one raised eyebrow and a twinkle in his eye.

"Let's take it from the top—don't get bit. Bit by mosquitos, that is."

Dr. Turner explained that the whole of Madagascar was considered a malaria risk zone. Malaria is caused by parasites that infect certain types of mosquitoes. In Madagascar, there are three different species of malaria parasites that infect humans: *Plasmodium falciparum*, *Plasmodium vivax*, and *Plasmodium ovale*. The worst of the three kinds is *Plasmodium falciparum*—it can cause severe and life-threatening malaria.[3]

"Anyone can get malaria," Dr. Turner told me with a serious tone. "All it takes is a mosquito bite. Do you expect to be spending a lot of time outside at dawn and dusk?"

I paused, knowing what I was about to say wasn't the correct answer. "Well . . . yes. We will be hiking and collecting field data, which means being in the forest before the sun comes up. I'll also be living out of a tent, and so that means all of our meals will be outside—likely around dusk."

Dr. Turner nodded, gravely. "Alright. Just make sure that you wear a long-sleeved shirt and long pants. And, of course, you'll have to go on some malaria medication while you are there. Let's talk choices."

Dr. Turner presented three different options for malaria medications. Malarone, he said, was the pill with the fewest side effects. You take it once a day, starting just a few days before you leave. Trouble is, Malarone is expensive. Most people who take Malarone, Dr. Turner said, are last-minute travelers going on vacation for a few weeks, so the cost isn't a big deal for them. But to purchase three months' worth of pills would cost hundreds of dollars, and we both knew that was a nonstarter—not on my graduate-student salary.

The second option was doxycycline—cheaper, but an antibiotic. Dr. Turner pointed out that it's not a great idea for women to take an antibiotic for months at a time—yeast infections waiting to happen. Another major drawback of doxycycline was that it increased sun sensitivity. I reiterated that I would be living and working outside, often in the hot sun, and so Dr. Turner presented the third option: Larium. Larium was within my budget but, as Dr. Turner noted, it is known to cause "minor hallucinations" and "weird dreams." He ascertained that I didn't have a history of depression or any other psychiatric conditions. Larium didn't sound ideal from a psychological perspective, but it seemed to be the best option for me, and Dr. Turner wrote up the prescription.

"Next up: don't get hit. What do you think that tip refers to?"

"Road safety?" I guessed.

"Exactly."

Dr. Turner explained how most tourists take the time to think about travel vaccines and prescription medication, but that one of the most hazardous elements of international travel is driving. In fact, he pointed out that motor vehicle crashes are the leading cause of death among healthy travelers.

"Make sure that you always wear a seat belt and avoid overcrowded or top-heavy buses."

I nodded gravely, but my mind flashed to something Shawn had described earlier. The *taxi brousse*, or "bush taxi." Shawn told me that these minibuses are a cheap alternative to renting a four-by-four. For a mere $5 Canadian, you could travel as far as 160 kilometers. Malagasy locals use the taxi brousse system all the time.

"It is a cultural experience," Shawn said. "You should definitely try it at least once. Just be warned: you get what you pay for."

The taxi brousses, Shawn said, are invariably jam-packed with people, their luggage precariously strapped—or just balanced unsecured—on the rooftops.

"On longer rides, the taxi brousse companies are pretty good about assigning one seat per person. But as distances get shorter, the buses will stop along the way to collect people from the roadside, squeezing in as many as possible. It's easy to get up close and personal very quickly with your fellow passengers." Shawn laughed to himself. "You won't believe what they pile on top of the vehicles. I've seen it all: furniture, food, bicycles, anything goes. More than once I have climbed into a taxi brousse and settled into my seat, only to look down and discover that I am sitting next to a basket of live chickens."

As Dr. Turner explained the dangers of overcrowded, overweight, or top-heavy buses, I wondered whether I should tell him about the taxi brousse. Would he be okay with that cultural experience? I decided to keep quiet as he continued.

"Watch carefully when you cross the street, and most importantly, avoid driving at night."

Okay, I thought, I can follow two out of four of Dr. Turner's road safety rules. That's not too bad . . . right?

"Last tip," Dr. Turner continued. "Don't get lit. In other words, be careful around alcohol."

Yup, I thought, gravely. Learned that lesson already—the hard way. I had made the rookie mistake of sampling some homemade cashew wine on a beautiful afternoon near the beginning of my master's research project in Belize. Travis and I and my master's supervisor, Mary Pavelka, had spent the day setting up project logistics for the howler monkey population survey we were about to commence. We had made our way to Monkey River Town to wind down at Ivan's, the local bar. Some American ex-pats had just passed through before we got there and had given a whole vat of homemade cashew wine to Ivan.

Someone poured me a glass of the sweet, cool drink, and it went down easy. A little too easy. Not only that, but the glass proved miraculously bottomless. By eight p.m., I was drunk as a skunk, and Travis had to carry me back to our cabin. I had never in my life been as sick as I was that night from alcohol, and I have never been that sick from it since. I know now that I had alcohol poisoning. Later, Travis told me, "That's when I knew we were meant to be together. You have never looked or smelled worse, and yet I still loved you, and even found you oddly attractive."

I suffered greatly for the next two days and made a mental note to self: do not drink homemade alcohol while traveling. Correction: do not drink homemade alcohol, full stop. Mary couldn't help but laugh at my misfortune. "It's a lesson we all have to learn the hard way, Keriann," she had said the next day, as I sat nursing some sparkling water. "That's why I opted for beer."

When I relayed the cashew wine story to Shawn in his office one day, he chuckled and told me about *tokogasy*, the Malagasy moonshine that is made from sugar cane. "Best avoided," he had said, and left it at that.

As if reading my mind, Dr. Turner reiterated, again, what I had learned in Belize: "Do not drink the local moonshine," he said firmly.

Oh, don't you worry, Dr. Turner, I thought. I'm right there with ya.

As I walked out of the clinic, clutching my prescriptions, it hit me. I'm actually doing this. It's happening. I rolled my left arm forward and then back. It was sore from my typhoid shot. The uncomfortable bruised feeling in my arm was hard evidence. Something tangible that I could feel. The dull pain throbbing in my deltoid muscle meant that all of this was real—in just a few weeks I would be more than 14,000 kilometers away from my home in Toronto. In the wilds of Madagascar.

> *I cannot stress the importance of batteries enough. Don't wait to purchase them in Madagascar because they will not be of as high quality as the batteries we can buy in Toronto, and they will cost twice as much. Batteries are especially important if you are going to conduct nocturnal lemur surveys. I'd suggest visiting Costco if you can, to buy your batteries in bulk.*

"Are you sure you need all these batteries?" My brother, Carlin, sounded doubtful as he plopped the double A, triple A, and C batteries onto the conveyor belt at the cashier. "What are you planning to use these for again?"

Carlin was the only person I knew who had a Costco membership, and so I had asked if he could take me to get my Madagascar supplies. He had quickly agreed—he loved going to Costco and loading up on food and other items in bulk. When we were growing up, Carlin was the outdoorsy one. Two years my senior, and often mistaken for my twin, he loved camping, mountain biking, and skiing, and he would always be the first one up on the day we embarked on a family excursion, helping my parents pack the car or the tent trailer. I, on the other hand—well, let's just say that I was *resistant* to the outdoors. My parents would have to pry me from my bed those mornings when we set out on some outdoor adventure. I enjoyed the activities once we got started, but for me, a preferred outdoor weekend activity involved riding my bike to the local convenience store, loading up on Archie comics and junk food, and spending the afternoon reading and snacking while lying in the hammock that hung in a shady spot of our suburban backyard.

"Oh hey," Carlin said, bringing me out of my daydreams and back to the Costco checkout. "Maybe you should go back and get some walkie-talkies? So you can communicate with your field assistants?"

"Way ahead of you," I said, lifting the heavily packaged Motorola two-way radios out of the cart.

I could see Carlin looked relieved by this purchase, and I knew immediately where his mind was: Panorama Mountain Resort in British Columbia. Carlin had been living in Montreal while attending university and had come back for a Christmas visit. We had gone skiing for the day with my parents. Carlin and I had decided to take off from my folks so we could hit up some more difficult runs. Carlin is an expert skier and snowboarder and I am not bad myself. Despite my resistance, our family did a lot of skiing, Calgary being so close to the Rockies. My parents spoiled us with yearly season passes and, early on, enrolled us in weekly lessons. Every weekend we would drive out to the mountains to ski. We would rotate seasons between Lake Louise and Sunshine. While Carlin and I are both excellent skiers and can handle most runs, Carlin was always better than

I was. He could handle the moguls on skis so gracefully that sometimes people watching him from the chair lift broke out into spontaneous applause. And on a snowboard, he could tackle deep powder like a pro. I was never that graceful.

The difference between our abilities didn't have much to do with skill—we had both taken the same lessons, after all. And I had even started at an earlier age and gone on to take additional lessons after he stopped. I should have been at least as good as he was, if not better. But I'm not, and I wasn't, and the difference, I've come to realize, is headspace. It's my hesitancy and fear—all in my head—that makes me a lesser skier.

Although we normally went to Sunshine or Lake Louise, that day, after hearing a favorable snow report, we headed to Panorama. Why not try somewhere new—it was only an extra hour to get there. I was feeling good. I was keeping up with Carlin, although he is a speed demon on that snowboard. We headed up to Taynton Bowl and made our way toward some black diamond runs. It was a snowy day, and we were dealing with deep powder. Powder skiing was one of my weak spots, but so far, I had managed. Carlin took the lead as we turned onto a black diamond run called "Stumbock's," which took us through some trees. Carlin was quick, and soon he was so far ahead that I couldn't see him anymore. I was struggling in the powder, and fearful of hitting one of the many trees that peppered the run. Although most skiers live for powder, I have always had a tough time with it. Skiing powder has always made me feel as though I am fighting against the snow. The snow wants my skis to move in one direction, even though I want to go the other way. There, on Stumbock's, I felt as though I was losing control of my skis—of the run. I stopped.

"Carlin? Wait up!" I called, hoping he could coach me down. My voice echoed, but I heard no response.

I came to a fork offering two directions. Which way had he gone?

"Carlin?" I called again. Nothing. And there was no one else in sight. I was alone. I decided to go left and hoped that I would meet Carlin at the bottom. But the run was steep and the powder so deep that a few meters in, I took a nasty spill, losing my skis and poles in the process. I clambered back up the hill to retrieve them. The steep incline and deep powder made it nearly impossible to clip back into my skis. After about ten minutes of trying, I sat down in the snow, defeated.

"Carlin?" I cried. But I knew he was long gone. He must have turned right.

Now what? I was stuck. There was no one in sight. Should I just wait until someone came along who could call the ski patrol to get me down? There it was: fear. Hesitation. I sat for a few minutes, quietly crying. I looked down at the steep run below me. I realized that no one was coming to help. I had to tackle the fear alone or risk sleeping on the mountain. And so, I held my skis and poles in one hand while I scooted down the hill on my butt.

It wasn't graceful. I wasn't proud of myself—but it worked. I managed to reach a less-steep area where I could clip into my skis. Carlin had our only map, so I had to rely on instinct to find my way down. Eventually, I connected to a blue run—Taynton Trail. From there, I knew I could find the lodge. Relief washed over me. When I arrived at the lodge, I hadn't seen Carlin for more than an hour. I was starting to take off my skis to go inside when he came running up, frantic.

"Keriann, are you okay? I was so worried. What happened? I ran into a ski patrol on my way down and he said you might be wrapped around a tree."

Sure enough, we had taken opposite directions at the fork. I had ended up on the Turnpike 1, a steep double black (a very advanced ski run), while Carlin had remained on Stumbock's to the end. He was really concerned. He thought he'd killed his little sister, after all. And he vowed to purchase handheld walkie-talkies for next time.

I was nineteen when that incident occurred. And, although it wasn't a life-and-death situation—I wasn't injured, just frightened—it taught me something about rising to challenges and helping myself. I grew up a lot that day. I realized that I needed to trust in my abilities and myself. Now, standing at the Costco checkout counter, I looked forward to seeing what else I could do.

"We won't have any electricity," I told Carlin, as we watched the cashier ring through the box after box of batteries. "And we'll be walking around looking for lemurs at night. See here?" I pulled out Shawn's guide and pointed to the battery chart. "It says I'll need at least forty double A's a week, five triple A's a week, and ten C's a week. In fact"—I paused to do some math—"I'd better run back and grab another sleeve of triple A's."

Carlin shook his head in disbelief: "Looking for lemurs in the forest at night? Are you sure you want to do this?"

I had to laugh at the look on his face. Even though Carlin had always been the outdoorsy one when we were kids, he had changed a lot since then. When we entered our twenties, we changed roles. Carlin began taking advantage of the city life. He had lived in Montreal while doing his undergraduate degree at McGill, and then moved to Toronto for his law degree. Over the span of those six years, he had taken to shopping at Holt Renfrew on a regular basis. Now, he was a civil litigation lawyer and had not gone camping in years. Meanwhile, I had remained in Calgary and met Travis—a die-hard outdoorsman—who had awakened the daredevil in me.

"Well, we'll be doing nocturnal surveys, if you want to get technical," I explained. "And yes, I am sure."

The plan was to set up a new research site in Kasijy, where Shawn and I and later other students could collect data on forest ecology. After we scouted out the locale and set up camp, we would cut line transects—straight-line trails through the forest—to enable us to look for lemurs. We would conduct population surveys, or counts, of all the individual lemurs that we spotted on our daily walks of the trails. These would include both diurnal and nocturnal species. From there, we would be able to calculate the total population density.

By cutting our transects perpendicular from the forest edge (the boundary between forested and non-forested habitat) into the interior of the forest, we would also be able to see which species preferred, were neutral to, or avoided the forest edge. Shawn had run similar studies in the eastern forests of Madagascar, at a site called Vohibola (voo-hee-boo-la), and found that the greater dwarf lemur had lower densities in edges, whereas the other five species found in the area either had higher densities in the edge habitats or didn't show any preferences between edges and interior habitats (thus were "edge tolerant").[4] Shawn deduced that the negative response exhibited by the greater dwarf lemur might be because this species is known to enter *torpor*, a state of reduced body temperature and metabolic rate that allows individuals to survive periods of reduced food availability.[5] The ambient temperatures are higher in edge habitats, and are known to inhibit torpor in other species of lemurs,[6] which could be why the dwarf lemurs avoid edge habitat. Also, the negative response could be related to variations in food abundance or predation pressures.[7] Shawn figured the other lemur species showed neutral or positive responses to the forest edge due to their diet. For example, the eastern woolly lemur eats mostly leaves, and leaf quality may be higher near forest edges because there is more light, which leads to higher protein concentrations in leaves.[8] Similarly, the brown mouse lemur eats mainly insects, which may be more abundant near the edge.[9] In Kasijy, we would see how the western forests and species of lemurs compared.

Understanding how habitat edges impact lemurs in Madagascar is crucial for conservation efforts because the country's forests are disappearing at an alarming rate. About half of Madagascar's forests have been converted to non-forest habitat since the 1950s.[10] Madagascar also experiences forest fragmentation. Fragments are best understood as islands of forest surrounded by "seas" of non-forest (often savannah habitat in Madagascar). More forest fragments means more forest edges. More edges can have devastating repercussions for lemurs, especially for those species that avoid edges. For these edge avoiders—like the greater dwarf lemurs—more edges means less habitat.

"Don't worry, it's just camping," I told Carlin, "with the added bonus of lemurs. You used to love camping, remember?"

He did not look convinced.

> *Make sure that you have purchased all the equipment that is needed to complete your project. Equipment will include camping gear as well as field equipment and cannot easily be purchased in Madagascar. You will also need to supply gear for your Malagasy student research assistants. I'd recommend buying two to three extra-large duffel bags to transport your luggage.*

After our Costco run, Carlin dropped me back at our parents' house in the Beaches area of Toronto. I had moved back in with my folks for a few months to save money while I completed the first year of my PhD program. It made sense. Travis wasn't living in Toronto yet, and rent was crazy expensive. I pitched in for food and was able to sock away money for future years.

Turned out, I would need it. Completing a PhD at the University of Toronto is an expensive proposition. At the time, students in the Anthropology Department would typically receive a funding package that consisted of about $14,000 plus tuition (another $6,000), which we earned working as teaching assistants for instructors in the department. Out of that funding, students had to cobble together rent, food, and other necessities like health expenses. Never mind luxuries like seeing movies or visiting the Graduate Student Union pub. To make it through the average five years that it takes to complete a PhD, many graduate students borrow substantially just to make ends meet. That is what I had to do, even though I came into the program debt free, had savings socked away from living with my parents, and received scholarships from the Ontario and Canadian governments throughout my program. In fact, sitting here now, eight years after finishing my PhD, I am *still* paying off my student loans.

Fortunately, the Anthropology Department partially funded my preliminary field research. With Shawn's help, I was able to secure $5,000 to pay for field equipment and my flight to Madagascar. And I spent every penny. I didn't own any camping gear—I had always relied on Travis for equipment—so I took a few trips to the outdoor store in downtown Toronto, Mountain Equipment Co-op, where I bought the necessities. I started with the big stuff: a new tent, sleeping bag, and sleeping pad. Because I would be going it alone, I opted for a two-man tent—just enough space for me and my gear. After some research on the internet, and a conversation with Travis, my own personal outdoor gear specialist, I chose an MSR "Hubba Hubba." Designed for backpackers, it is a three-season ultralight tent, easy to put up and down, and it packs small. Plus, what a fantastic name! Say it again . . . *Hubba Hubba.* Travis had also suggested that I go for a mid-weight sleeping bag.

"It's always colder than you think it will be," his words echoed in my head as I pulled my selected bag off the top shelf and tucked it under my arm.

As for the sleeping pad, I had taken a suggestion from Shawn and not skimped on the size.

"I'd get the biggest Therm-a-Rest pad they have," he had advised. "You'll be there for three months; it's worth it."

Whoosh! There goes $1,000.

On my second trip to MEC, my mom tagged along for support and to provide some much-needed carrying arms.

"Grab a basket," I directed when we entered the store. "I'll take one too. I will probably fill these up quick. There's still a lot on my list."

We each grabbed a black rolling shopping basket from the stack at the entrance to the store. My mom diligently followed me through the various sections of the store as I crossed things off my list. I chose three extra-large cloth duffel bags for starters; I would need something to carry all my gear from Toronto to Madagascar. I added new hiking shoes, quick-dry pants, shirts, shorts, and socks galore. One could never have too many socks in the field—the flooding Belizean forests had taught me that. I also needed a hat.

"Here. Try this one." My mom helpfully handed me a large-brimmed, UPF 50+ outdoor hat.

"Here goes," I said, and put on the hat. I couldn't help but laugh when I looked in the mirror. I already have a smaller-than-average head size, and the hat only accentuated that feature. I looked to my mother, who was smiling encouragingly.

"It looks like a good hat," she offered, no doubt reading the hesitancy in my expression. "It's not a fashion show, remember."

I looked back at myself in the mirror. I shrugged. I tried to remember if I had seen any photos of Jane Goodall in such a hat. I seemed to recall one photo . . . maybe. I reminded myself that I was heading to the northwest of Madagascar, where it would be hot. Very hot. The dry deciduous forests, I knew, had daytime temperatures in the dry season (when I would be there) that commonly exceed thirty degrees Celsius. Dang it. Yes, I was sure I had seen Jane in a wide-brimmed hat, I thought. Positive. I took off the hat and placed it in my basket.

Next I moved on to backcountry medical supplies. I grabbed a large, bright red zippered first aid kit, which I would later fill with bandages, gauze, scissors, blister protection, Polysporin, emergency ice packs, slings—the works. We would be hiking through remote areas of Madagascar. Though I hoped no one would get injured, cuts and scrapes at least were inevitable, and I knew that I had to be prepared for the worst. There would be no doctors where I was headed.

On Shawn's suggestion, I also picked up a few backpacking comforts: a solar shower and a pack hammock. I grabbed a couple of tarps, a compass . . . oh, and rope. Don't forget the rope.

"You never know what you might need it for," Travis had advised. "I usually buy my rope from the climbing section at MEC."

I led the way to the back of the store, where I could see the climbing gear—a wall of colorful climbing shoes, climbing harnesses hanging from racks, and, yes, behind a small desk, the various rope options coiled tightly around cardboard spools. Behind the desk was a slim blond guy hunched over, leaning on the desk by his elbows. I had never bought rope before, so I stood back and surveyed the choices.

"Can I help you?" the guy asked, not moving from his relaxed position, his eyebrows slightly raised. The tag hanging from his neck said his name was Mark.

"I need some rope," I said, stating the obvious.

Silence. Wow, this Mark fellow wasn't exactly welcoming.

I shrugged. "I guess I'll take some of the eight-millimeter cord."

Mark sighed. "How much do ya want?"

I thought about it, running some math in my head. "Let's go with . . . ten meters."

"You're not climbing with this, I hope." Mark sighed again. I could see what he was thinking. The skepticism and annoyance was written all over his face. Mark had given me one look and made his judgment. I was just some schmuck who didn't even know what kind of rope was the good rope.

Here's what I wish I had said: "Mark, is it? I am taking that rope with me to Madagascar. That's right . . . Madagascar. I'll bet you don't even know where that is. I'll be using that rope at my remote field site in the deep forest, where I will spend my days hiking and searching for lemurs. I'll be using the data that I collect there—in the wilds of Madagascar—to inform research papers that I later publish. So, no, Mark, I am not climbing with that rope, but I do need that rope, so be a sport and measure it out for me."

Instead, I just shook my head no, and waited as he slowly measured out ten meters of rope, which I then added to my cart.

"Let's go, Mom," I said quietly.

As we turned to go, she said, "You should've told him what you're using the rope for, Keriann. You're so much tougher than he is."

"Shhh! Mom don't worry about it," I whispered, glancing around to make sure no one had heard her.

"You always have been tough, you know," she continued, undeterred. "When you were two, I could not get you to stay in your stroller—even though you had only just learned to walk. You had the strongest little legs—I couldn't believe it—and you would just go, go, go, chasing after your brother." She paused and patted my shoulder. "You were—and are—one tough cookie—don't you forget it."

I laughed and shook my head. Maybe she was right.

My mom's encouraging words ran through my mind again when I later stared down at the rope laid out on the floor of my parent's finished basement, along with all my other gear. Having just returned from Costco, I added the batteries, books, and toiletries to the ever-growing pile on the floor. I surveyed my collection. Shawn had taken care of most of the data collection equipment, including handheld GPSs (global positioning systems), range finders, measuring tape, and compasses, all necessary items for creating line-transects and running lemur surveys. I had also bought three boxes of brightly colored vinyl flagging tape to mark the line-transects, several waterproof field notebooks, my binoculars—arguably the most important of all tools for a primate researcher—a camera, a headlamp, and an assortment of other camping and research supplies.

Another important piece of equipment was my water filtration system. Kasijy is located near the roaring Mahavavy River, so there would be ample water available at the site. But we couldn't drink it as is—it had to be purified. In Belize, we had purified our water with treatment drops. For Madagascar, Shawn recommended I bring along a water pump in addition to the treatment drops, and so on one of my many MEC visits, I had purchased an MSR water purifier, complete with an attachment for my one-liter Nalgene bottle. I had already taken it out of the cardboard and had a go with it over my parent's sink. Step one: attach hoses. The black hose, attached to the filter in the dirty water, the clear hose affixed to the Nalgene bottle attachment. Step two: attach float to the dirty-water hose. Step three: pump like you have never pumped before. The directions had stated that I should pump two liters of water through the pump before using, to remove any residual carbon dust. Okay, then. As I pumped the water, which had come from the Toronto tap, I thought about how the next time I used the pump I would likely be pumping water from the Mahavavy River, which ran adjacent to Kasijy. I wondered if that water would flow as easily through my water filter as the Toronto tap water. What would it taste like?

The water filtration system wasn't the only piece of equipment I tested. I also went through the exercise of setting up my tent to make sure all the bits and pieces were accounted for; I blew up my Therm-a-Rest and let it sit out for a night, to make sure there weren't any holes; and I wore my hiking boots around Toronto to break them in—quite a fashion statement, I must say. I was ready.

Beyond the equipment, I had learned in Belize to bring entertainment. Fieldwork involves a lot of hard physical labor, but it also delivers a lot of down time. And this trip, without Travis, was certainly going to involve a lot more alone time than I was used to. With that in mind, I selected eight books and purchased a video iPod (cutting-edge technology back in the day), which I planned to load up with episodes of *Gilmore Girls* and a few movies. On Shawn's advice, I also picked up some of my favorite candies, Sour Cherry Blasters, for the most desperate moments. Finally, I gathered some items to do a week of traveling around Madagascar after I finished my fieldwork. I bought a Madagascar guidebook and hoped to travel to a few other areas of the country to check out different kinds of forests and lemurs while I was there. As I stood surveying my bounty, my dad knocked on the door.

"Knock, knock," he called down from the kitchen upstairs.

"Come on down," I replied.

"I just wanted to see what you've put together here. Wow! How do you figure you are going to carry all this stuff?"

I picked up one of the extra-large MEC duffel bags that I'd bought. "Well, I have three of these. I'm technically only allowed two bags, but I'm going to pay to bring one extra. I need to bring enough books, ya know?"

He nodded. No argument there. My father has published more than a dozen books and is quite well known as an author. He rubbed his black beard, which he'd had since before I was born, and continued to stare at the piles of stuff and nodded, a distant look in his eyes. "We're going to miss you," he said, putting his arm around me. "You be safe out there. I'm not sure where you got this travel bug."

"Dad!" I rolled my eyes and made a face.

My father was always slightly overprotective of me, his only daughter—a trait that I've never been able to reconcile with his own adventurous past. In his youth, he'd hitchhiked across the country from Montreal to San Francisco, where he worked as a bicycle messenger—back in the days before helmets! After he and my mom met, they had lived in Tanzania, where they'd worked as teachers. Before that, the two of them had spent one summer living atop a mountain in the Rockies, working as fire lookouts. So, standing there in front of my field gear, as I was about to depart for Madagascar, we both knew where I had got my lust for adventure.

"Alright." My dad clapped his hands: "Let's get you packed."

We spent the next hour packing gear and trying to distribute the weight evenly. The batteries, especially, were weighing things down. Luckily, I was something of an expert at pushing luggage weight limits. When we went to Belize, we would bring a lot of dried food, including heavy energy bars, and I would always manage to sneak in just under the weight restrictions. I would weigh myself alone on my parents' old analog bathroom scale, then weigh myself again holding the bag and do the math. The results were surprisingly accurate. For this trip, I could bring two checked bags that each weighed less than fifty pounds. I also planned to pay to bring one extra fifty-pounder. I zipped up the last bag and turned to my dad. "Okay," I puffed. Packing always makes me sweaty and aggravated: "I think we've done it. Too bad I can't bring all the books, but the batteries and medical supplies are more important."

"More important than the books?" My father—the author—gasped. "The mind reels."

———————————

Email from Keriann McGoogan to Travis Steffens on May 15, 2006
Re: hi
Well, I can't believe that that is maybe the last time I will talk to you for several weeks at least. Hopefully we can arrange to Skype occasionally. . . . I think that is going to be a must. I miss you already!!

I hope everything is going well for you. I am still working diligently on getting stuff onto my iPod. I just downloaded the engagement photos so that I can look at those when I miss you. At least that will comfort me. And on the iPod, I can set the slide shows to music! So my dream will come true after all.

Tomorrow I drive out to Peterborough with my folks to visit my granny. Then more iPod stuff. On Wednesday I will run with my neighbor and then Shawn's wife is dropping off some stuff he forgot to bring. Hopefully it will fit in my bag.

———————————

Part II

Toronto, Canada, to Antananarivo, Madagascar.
Map created by Travis Steffens.

CHAPTER 3

Up, Up, and Away

May 20, 2006

On Saturday, May 20, my flight to Madagascar left at six p.m. I would arrive in Antananarivo near midnight on May 22. I had spent the day leading up to my flight at my parents' house, packing up the last of my gear, futzing with my carry-on luggage, stretching my legs with a jog, and just trying to relax. This would be the longest flight I had ever taken. I would spend seven hours flying from Toronto to Paris. Then, I would hop on a connecting flight from Paris to Antananarivo, which would take ten hours. Before this, the farthest I had flown was to Europe, so what happened on flights beyond seven or eight hours remained a mystery to me. I had talked on the phone with Travis a few days before, but the connection hadn't been great. He was in Belize and on his way to Rio Bravo to check for howler monkeys.

I confessed to Travis that I was feeling nervous. "I realized yesterday—I'm going to be the only woman on this trip!"

"Really? That's pretty cool. You're hardcore."

The original plan did not involve me spending time alone in Madagascar. When I first started my PhD at the University of Toronto, Shawn had another PhD student, Hilary, who had entered the program that same year. I knew her from the master's program at the University of Calgary. We had gone through the first semester of our PhD together—classes, grant applications, proposal writing—and she was supposed to join me on this preliminary trip to Madagascar. However, at the beginning of the second semester of the program, she decided to transfer to a different program at a different school.

Hilary broke the news shortly after the Christmas break. She was a former dancer-turned-yoga enthusiast and looked the part. Tall, slim, with dark brown hair, she sat with perfect posture when she told me: she had decided to move back to Alberta with her long-term boyfriend and to pursue biology, a better fit for her, she explained. Up until that point, I'd had a perfect picture in my mind of the two of us spending the three months in the Madagascar wilds together—marching through the forest together, hanging out by the fire after a tough day in the field, navigating the inevitable hardships, but also making the fun memories. But now, well, it only took a moment for that mental picture to disappear in a cloud of smoke. *Poof.* Hilary wouldn't be coming to Madagascar with me. I would have to go it alone for much of the trip.

Sure, Shawn would be with me for the first while, but he would not be staying the whole time. He and I would overlap for about a month, which left me with about six weeks working in the field on my own. I kept reminding myself that I would not be completely alone. I would have two Malagasy research assistants from the University of Antananarivo, and we would hire a few local villagers as forest guides when we reached Kasijy. But I would be the only Westerner among us. And I would be the only female.

Knowing that I would be the only woman on the trip was both a point of pride and a scary thought. On the one hand, I relished the opportunity to show the world what women can do. I had already toughed it out in the harsh Belizean forests—even wielded a machete—and now I would be going it alone in the wilds of Madagascar. How cool was that? But I also had a niggling worry about the challenges I was about to face as the only woman on this trip among a group of men. As an exercise, try this: Google "Why is it more difficult for women to travel alone?" and see how many hits you get. I got over forty million—articles from around the world exploring the "uncomfortable truths" about traveling solo as a woman, and "why travel safety is different for a woman." Like it or not, female travelers face challenging realities such as navigating cultural norms attached to being female, the unwanted attention that we receive, being viewed as more vulnerable, and worst of all, the possibility of sexual harassment. I worried, too, that I would be automatically viewed as less strong and less capable than the rest—that I would have to prove myself and work extra hard to earn the respect of my team members.

I gave Travis the details on how to reach me in Tana at the hotel and via the satellite phone when we were camping: "You can send me messages on the satellite phone that are 160 characters maximum. I probably won't be able to respond much, but at least it's something."

After we'd said our tearful "I love yous" and signed off, it hit me again. I was really doing this. In a few days, I would be in Madagascar, chasing lemurs.

Arrive at the airport at least 3–4 hours before your flight departs from Toronto. Make sure that Air France sends your luggage through to Madagascar.

Although it was only twenty-seven kilometers from my parents' house in the Beaches, the journey to Pearson International Airport was daunting. It is at least a forty-five-minute drive each way in what is usually rough traffic. My father stepped up to the task. My mother was distressed to learn that there would not be room for her in the car.

"Are you sure I can't come, Ken?" My mother didn't like the idea of missing out on the traditional goodbye hug at the airport.

My dad was ready to be persuaded, but then came the problem of my three massive duffel bags. Did I mention that they were extra-large? At 140 liters, and each weighing almost exactly fifty pounds, they were simply too large for there to be enough room for three people in my parents' Honda Accord. The decision had been made for us. I said my goodbyes to my mother and my beloved cocker spaniel, Cody, at the house, and hopped into the car.

I sat in silence with my father as we navigated the highway to the airport. It was a comfortable silence, both of us somewhere else in our minds. When I was growing up, this kind of behavior would drive my mother crazy. She could not believe that the two of us could just sit together in silence, in our heads, navigating our own internal dialogues, for hours on end.

"Why am I the only one talking?" she would often cry out. "Are you two listening to me?" (Sorry, Mom!)

I am not sure where my father was, but I was in Madagascar. In my mind's eye, I was already in Kasijy. Deep in the forest, watching the lemurs. I could see them. The snowy-white sifaka, with its powerful hind limbs clinging onto a tree trunk. When they move, they rapidly extend their powerful hind limbs and propel themselves upward and outward from one vertical support to another.[1] That's called vertical clinging and leaping. And then I could hear their call. Sifakas get their name from a characteristic alarm call behavior in which they exclaim a vocalization that sounds like "shee-*faak*"—reminiscent of a sneeze— and rapidly jerk their head backward several times.[2] In my mind's eye, I could also see the diminutive mouse lemur, small enough to fit into the palm of my hand, popping in and out of the thick foliage. The mouse lemur moves quadrupedally—on all fours—and sleeps during the day in nests of dead leaves or tree holes.[3] And wait—was that a chameleon? My mind raced as we wound our way to the airport. I glanced over at my father. Suddenly I knew exactly where his mind was.

"Are you crying?"

He shook his head but didn't respond.

"Don't worry, Dad!" I said. "I'll be back soon."

He nodded and smiled. "I know that. It's just . . . you're my little girl. And you're heading halfway across the world."

Did I mention that my father wears his heart on his sleeve? I do love that about him. I remember one occasion where I was watching a film in my parents' basement—*Jane Eyre*. My father popped in briefly and caught the last fifteen minutes. Even though he hadn't watched the whole movie, he still choked up as we watched Jane find Rochester blinded, Thornfield having burned to the ground. If it took only fifteen minutes for him to become emotionally invested in Jane and Rochester's story, it was of course no surprise now that he was struggling as he sent his only daughter and youngest child to the far side of the globe.

"But it's just a few weeks, really, Dad. And Shawn will be there when I arrive. I'm excited." I paused, but then confessed: "I am a *little* nervous."

We arrived at the airport and pulled into the parking garage. I hopped out of the car and grabbed a couple of luggage carts. My dad helped me load the bags onto them. Onward we rolled into the terminal and over toward the Air France check-in counter at the far end. I could see their familiar white and blue logo. It was calling me.

But as we neared the counter, I noticed that it was eerily quiet. No one was behind the desk. Was I in the right place? I glanced around, fighting anxiety. Behind me, a man and a woman with their two children were sitting in a row of chairs, their luggage carts—piled higher than mine, believe it or not—in front of them.

"It's not open yet," the woman said gently when I glanced in her direction. "They told us it will be open in twenty minutes."

"Thanks," I said, relieved. It was still a good four hours before my flight was to depart, so it was no surprise that we were too early for check-in. But, better too early than too late, I thought.

> *You can pay for your excess baggage and then deposit your luggage at the oversized counter.*

"What's that?" The airport security guard at the oversized baggage scanner pointed to the black-and-white image in front of him. I squinted at the screen.

"Those are batteries." I hadn't taken the batteries out of their packaging and the way they lined up in my bag made them look like bullet cartridges. Whoops.

"Why do you have so many batteries?" the guard asked.

"I'm going to be camping and hiking at night," I replied, my heart beating more quickly now. I was starting to sweat. Even though I didn't have anything illegal in my bags, being stopped by airport security always causes me some anxiety. Should I tell him about the fork-marked lemurs I might see during the nocturnal surveys? I doubted he knew about these creatures and their distinct locomotor style. Fork-marked lemurs are known to break into a fast run along horizontal branches, jumping from branch to branch without pause. When they stop, this lemur bobs and wags its head, creating a distinctive eye-shine that I could view by using my battery-operated flashlights.[4] If he knew all of that, would he forget about the batteries?

The guard proceeded to open my bag and poke around.

"Okay," he said, and I breathed a sigh of relief.

I watched as my three large duffel bags chugged along on the conveyer belt and disappeared, the first step in their journey to Madagascar. I turned back to where my father was waiting for me. I was now carrying only my backpack and a small travel purse.

"All set?" my dad asked.

"Yup," I replied. "Fingers crossed those bags make it the whole way." I had heard horror stories from friends about lost luggage—lost field equipment. Shawn had warned me before I left: "Many a bag goes missing en route to Madagascar. Don't panic if yours doesn't show up right away. Just try and distribute your equipment evenly so that it won't be the end of the world. It's a good idea to include a spare set of clothing and your most important gear—like your binoculars and computer—in your carry-on luggage. Oh, and on the way back, make sure you make copies of your data sheets."

My father and I looked at each other, knowing that it was time now for the tough goodbye. I could see he was again fighting back tears. I hugged him tightly.

"You take care of yourself," he managed as we pulled apart.

I turned toward the security line. As I walked forward, my face flushed and I could feel the tears brimming. I couldn't bring myself to look back.

Be sure to catch some zzz's on the seven-hour flight from Toronto to Paris. Once you arrive in Paris, you will need to deplane on the tarmac, board a shuttle bus, and make your way to the terminal with the rest of the passengers. Hopefully, you will have secured a seat near the front of the plane so that you will be able to board the first bus. Be aware connecting flights between Toronto and Paris are quite often very tight, so move as quickly as you can.

I stood trying to stabilize myself on the moving bus at the Paris airport, wearing my backpack and travel purse and clutching my passport in one hand while hanging on to the handle above with the other. I was surrounded by fellow passengers who had crammed onto this first bus to meet connecting flights. I had about one and a half hours, which meant I had to hustle. But living in Toronto and frequenting the TTC—Toronto's packed public transit system—had pre-pared me well, and I had elbowed my way into a decent position.

We arrived at my terminal, and I followed several other passengers into the building. I scanned the area for signage and spotted an electronic list of flights, departure times, and gates. I noted my gate and then rode an elevator to the security check. Since I had checked my luggage through to Madagascar, I had nothing to collect. This luxury contrasted starkly with how things had gone when I flew to Belize via Houston. The rigorous American security system meant that you had to pick up your luggage at the connecting airport, which added an extra layer of stress to the travel experience.

After taking a few wrong turns, I found my gate. A small feat, perhaps, but this was the first time I had navigated a foreign airport by myself. I am not known for being great with directions (insert knowing guffaw here from friends and family). Having arrived with plenty of time to spare before boarding the plane to my dream destination, I decided to pop into a café and treat myself to an Orangina and a croissant—this was Paris after all—and then sat down to wait for boarding. Through the huge window at the gate, I could see the sun peeking out from behind the passenger planes lined up waiting to take people up, up, and away.

After you board your flight to Madagascar, settle in for an 11–12-hour flight. Try to sleep, get up and walk around a lot, and watch some movies.

I had a window seat on the flight to Madagascar. Beside me, a small, middle-aged French woman sat down and immediately began working on her laptop. Not a bad seat companion for a long flight, I thought, peering a few rows ahead to where some large, sweaty men had taken their seats. I gazed out the window at the tarmac. It was hard to believe that when we landed I would be in Antananarivo. I turned to my reading material. My trusted Conservation International field guide to the lemurs of Madagascar details one of the earliest and most famous accounts of Madagascar and lemurs, written in 1648 by the French merchant Etienne de Flacourt, the director of the Compagnie Française de l'Orient.[5] He was charged with revitalizing the trading posts of Fort Dauphin, found at the southern tip of Madagascar. His commercial efforts failed, but Flacourt made history with his detailed descriptions of what he found on the island. He stayed in Madagascar until 1655 and visited the southern and east-ern parts of the island, documenting the flora, fauna, and culture of the island. Flacourt famously described at least eight species of lemur, including one that is now extinct.

Flacourt writes about a creature he called the *tretretre* or *tratratra*, "An animal as big as a two-year-old calf, with a round head and a human face: the front feet are monkey like, and the rear ones as well. It has frizzy hair, a short tail and human-like ears." He goes on to say, "It is a very solitary animal; the local people fear it greatly and flee from it as it does from them."[6] Many scientists today believe that Flacourt's writings could represent the only existing eyewitness account of *Megaladapis*, an extinct, gorilla-sized lemur.

I sighed to think that these giant lemurs existed well into the last millennium.[7] I had just missed them! The early explorers would have found an additional seventeen lemur species that have since gone extinct, and almost all of them would have been larger-bodied than any alive today.[8] Just like the living lemurs, the giant, extinct lemurs occupied a variety of niches, displayed a range of locomotor behaviors, and had distinct diets.[9] The largest of the bunch was *Palaeopropithecidae*, nicknamed the sloth lemur because of similarities to arboreal sloths. Researchers think that sloth lemurs were specialized hangers and that they fed on leaves, fruit, and seeds.[10] *Megaladapidae*, otherwise known as koala lemurs, were about the size of a female gorilla and thought to be committed tree dwellers—slow climbers—despite their large body size.[11] Then there was *Archaeolemuridae*, the monkey lemurs, who were among the last giant lemurs to become extinct. They are thought to have been the most terrestrial (ground-dwelling) of the giant lemurs, and able to break open hard objects like nuts with their teeth.[12] Researchers have found evidence of butchery marks on the bones of the giant lemurs, which indicates that they were hunted and eaten by humans in Madagascar.[13] Hunting pressure, combined with natural aridification, fires, and habitat disturbance are all possible contributors to the extinction of the giant lemurs in Madagascar.[14] Recently, though, researchers have further refined our understanding of why these giant lemurs disappeared based on radiocarbon records of subfossil vertebrates, new data on human butchery of the giant lemurs, and discoveries of cave deposits allowing for a reconstruction of climate.[15] The "Subsistence Shift Hypothesis" attributes the decline in Madagascar's megafauna to the expansion of the Indian Ocean trade network, and a transition from hunting and foraging to herding and farming. The trigger to the extinction event, these researchers argue, likely took place between 700 and 900 CE and coincided with an expansion in the human population and new settlers planting crops and domesticating animals like cattle.

During that expansion period, humans would also have hunted wild animals, including lemurs.[16] Larger-bodied species—like the giant lemurs—would have been more vulnerable to human hunting, required more habitat area to maintain a population, and also been slower to reproduce.[17] Knowing what I did about the precarious conservation position of the living lemur species, many of which are threatened with extinction due to habitat loss, hunting, and the pet trade,[18] I was grateful that I was heading for Madagascar before it was too late.

I rested my head against the airplane window and closed my eyes. As I drifted off to sleep, I could hear Shawn's final words to me in his office that April day. "This is an exciting time for you, Keriann," he had said as I stood up to leave. "You only get to experience your first visit to Madagascar once. And I can tell you from my own experience and from watching my students: Madagascar is a place that has a way of staying with you. I know you are going for the lemurs. And the lemurs are amazing. But you'll see that the Malagasy people and culture are equally incredible." Shawn looked off into the distance, "Madagascar is the type of place that compels you to go back again and again. It gets into your blood."

You will probably arrive at Ivato Airport in Anatananarivo at night. After you deplane, go down some metal stairs and onto the tarmac, just follow the hordes of people into the terminal.

When the plane touched down in Tana it was dark. I shielded my eyes and peered out the window anyway, just in case there was something to see. Nothing. The flight had felt long, even though I had managed to sleep for a good four hours. I had then watched two movies and chowed down on the two airplane meals—including baguettes and red wine in true French style. I had flipped through my Madagascar guidebook, read a magazine, and written in my journal to pass the time. My seatmate seemed to have a bladder of steel and hadn't budged from her seat through the entire flight. Turned out she wasn't the ideal seatmate. As I crawled past her to use the washroom for the third time, I vowed to request an aisle seat next time.

I surveyed the other passengers who stood in the aisle, anxiously waiting for the plane doors to open. Now that we had arrived, I found myself speculating about who the other passengers really were. Two rows in front of me sat a Malagasy family, a married couple and their two small children. Maybe the father was a schoolteacher, and they were all heading home after visiting cousins who had moved to Paris four years ago. And my seatmate? Suddenly, she was a journalist, working for Reuters in Madagascar, heading home to her house in the capital after a short visit with her family in France, where she'd met her baby nephew for the first time. Did any of the other passengers guess my story? That I was on my way to Kasijy—to chase lemurs in the wilds?

I carefully navigated the metal stairs of the plane and crossed the tarmac with my fellow passengers. The air felt crisp and cool, refreshing after being on a plane for more than ten hours. I followed the crowd into the airport terminal—and stepped into a total zoo. It was hard to believe that these hundreds of people had all come off one flight, but they had. We had been on a Boeing 777, the world's largest twinjet airplane, with seating capacity for roughly 396 passengers. Shawn's unofficial guide said the booth to purchase your three-month tourist visa would be on the right. I stood on my toes and leveraged my height for a quick scan above the sea of heads. About twenty meters away, I spotted a tiny booth occupied by a man dressed in a police uniform. I edged over to the end of a slow-moving line. After about twenty minutes, I finally reached the booth and slid the necessary documents through the small hole in the plastic window: my disembarkation/embarkation card, return flight ticket, a dozen euros, and passport.

As I watched the officer handle my documents, the reason for the slowness of the line became clear. The officer calmly and deliberately licked his thumb and forefinger and plucked a Post-it-sized piece of paper from a stack of identical papers. The paper, I would later see, had an image of the Malagasy flag (in vibrant white, red, and green) in the center, and read "Visa de Sejour à Madagascar" in block letters at the top. The officer picked up his pen and, in neat, compact handwriting, copied my name, passport number, and other details onto the piece of paper. Next, he flipped the paper over and, using a small brush, coated it in glue. He affixed the paper into my passport, where it covered an entire page, and finally took out a circular rubber stamp, slowly and firmly pressed the stamp onto the red ink pad, and then stamped the bottom left of my passport page with the seal for the Republic of Madagascar. It was a circle with an outline of Madagascar at the center, around which the sun's rays emanated and below which appeared the head of a zebu (a Malagasy cow). Pretty elaborate, I thought, as the officer passed back my documents. But a colorful visa is every world-traveler's dream—a badge of honor.

I made my way to the luggage carousel. All the bags had already been offloaded and the carousel had stopped moving. I grabbed a trolley from the side of the room and wandered around among the bags lined up on the floor. One, two, . . . relief washed over me when I saw that third bag. I loaded my luggage onto the cart and made my way toward the exit marked "Nothing to Declare," where I had arranged for a MICET driver to pick me up. MICET (Madagascar Institut pour la Conservation des Ecosystèmes Tropicaux) is a nongovernmental organization that works closely with researchers who come through Madagascar. An invaluable resource, MICET aids researchers in obtaining necessary permits, helps to translate research reports, provides training for Malagasy and international students, offers transportation to and from field sites, and collects researchers when they arrive at the airport. I would pay a very reasonable $300 facilitation fee for these services, and it would prove to be well worth it.

"Vazaha!"

As I approached the exit, I heard several voices, echoing around me. In seconds, three slim Malagasy men, wearing neon orange vests, surrounded me.

"Vazaha," they said again and reached for my cart. Shawn had warned me about this.

"As you exit the airport," he had said, "you will be approached by porters, yelling *vaa-zaa*, which just means foreigner in Malagasy. It's fine to use them to help you handle your bags out of the airport, but you'll have to pay them each a dollar US per bag. Also, just be aware that sometimes more than one porter will find a way to tag along and they will all expect individual payment. Just pay the one who helps you."

I turned toward the three men and held up my right index finger. "Just one," I said in my best French. "Une seulement. I only need one of you."

In the end, four of us exited the terminal together, one porter pushing the trolley and two resting their hands on the stacked luggage. I guess that was helping? I couldn't help feeling a little defeated. I scanned the sea of Malagasy faces behind the barricade and spotted a sign with a close approximation of my name written on it: "Keian." I made eye contact with the kind-looking and very small Malagasy man holding the sign and he waved and came toward me.

"MICET?" I asked.

"Oui," he replied. "I am Pierre," he said in a thick French accent, and we shook hands. Shawn had joked earlier about the typical Malagasy handshake. "The ol' limp fish," he'd laughed. That handshake with Pierre certainly qualified. I found myself dialing back on the firm North American–style grip that I had worked so hard to perfect.

It was hard to tell whether Pierre was shy or simply didn't speak much English, but we walked quickly and quietly, my three porters in tow. Outside the terminal we reached the parking lot and stopped at a blue minivan with the letters "ICTE" on the side. ICTE stands for the Institute for the Conservation of Tropical Environments—the American partner-organization of MICET—and was established by lemur-researcher royalty, and one of my academic idols, Dr. Patricia "Pat" Wright, who worked out of Stony Brook University.

As a female primatologist studying lemurs, I had drawn inspiration from Patricia Wright and her work in Madagascar. At that time in 2006, I had not met her in person, but I had referred to her research while preparing my PhD proposal and had been captivated by her talks at several primatology conferences. She had been working in Madagascar since the 1980s.[19] She was best known for the discovery of a new species of lemur, the golden bamboo lemur. The golden bamboo lemur has a unique diet of about 90 percent bamboo, and mostly giant bamboo (*Cathariostachys madagascariensis*).[20] The extremely high levels of cyanide found in giant bamboo would be lethal for any other mammal, but the golden bamboo lemurs are able to tolerate them,[21] possibly by consuming foods with higher protein, which can be used for detoxification.[22] Sadly, the golden bamboo lemur is listed as critically endangered on the IUCN Red

List of Threatened Species, with only about 630 individuals remaining in the wild—and their unique diet is a factor in their declining population.[23] People engage in slash-and-burn agriculture, and harvest bamboo—their primary food source—for building houses, carrying water, and making baskets.[24] To protect the golden bamboo lemur, Dr. Wright established one of Madagascar's most famous national parks, Ranomafana.[25] Years later, I would see her work featured as part of IMAX's *Island of Lemurs: Madagascar*. But now, as I watched the porters help Pierre load my luggage into the back of the MICET van—a van that existed because of Dr. Wright—I felt a connection with my fellow female conservationist and passionate lemur researcher.

Strong female role models are characteristic of the field of primatology. They include not just Pat Wright, but Jane Goodall, Alison Jolly, Birute Galdikas, Jeanne Altmann, and Linda Fedigan, to name just a few. In a 1994 article, Fedigan explored the unusually high numbers of female primatologists, and found that there are significantly more women in the field than in general biology, and more women studying primates than other types of organisms.[26] Fedigan suggested a number of possible explanations for this, one of them being a tradition of strong female role models.

Fedigan argues against one of the most common explanations for the pervasiveness of female primatologists, "the big brown eyes hypothesis." This theory suggests that women study primates because they are cute. Fedigan rightly points out that many primates are far from cute when you study them closely. Chimpanzees, for example, can be extremely violent, and regularly hunt and kill other monkeys.[27] There is also the possibility that women are attracted to primatology because of the biological nature of the primates themselves. Many primates are female-bonded, which helps direct focus on female animals, and may attract women to the discipline.[28] Lemurs, particularly, are distinguished from other primates because they show female dominance in groups.[29] Take the ring-tailed lemur, for example. In that species, females lead their groups from food site to food site, receive priority access to food resources, and will show direct aggression toward males.[30] Perhaps, as Fedigan argues, women are attracted to primates because female primates—like lemurs—are important, or successful, within their own societies.[31] Speaking for myself, I had been drawn to primatology for a combination of reasons, including a fascination with understanding humans and where we come from; an interest in primate behavior; and yes, the number of female role models, like Pat Wright.

Pierre waited while I paid the three porters. I managed to negotiate the price down from $3 US per bag each to $2 US per bag each—a small victory. "Misaotra," they said after I handed them their money (pronounced "mee-sew-tra," and meaning "thank you"). I hopped into the back of the van and glanced at my large men's digital Ironman watch (an unattractive accessory but a useful tool for fieldwork). It was nearly one in the morning. Pierre expertly navigated his way out of the airport parking lot and into the street. He knew where I was staying—La Maison du Pyla, a bed-and-breakfast-style hotel situated in the Tsiadana ("see-a-da-na") district of Tana, near the University of Antananarivo. Shawn had stayed there many times and knew the owner, Fanja ("Fon-za"), very well.

"Fanja is absolutely wonderful," Shawn had said. "She is a wealth of information on everything that you will need in Tana, and Pyla is clean and safe. I stay there because it feels like home."

Home, I thought, as we drove through the darkened streets. Home felt so far away right now, even though I had only just arrived in Madagascar. As we drove, I didn't see much. The streets were devoid of people and only a few vehicles shared the road with us. As we passed an area with several closed-up storefronts, I leaned forward. "Is this downtown?" I asked Pierre in French.

He looked back at me through the rearview mirror, brow furrowed. "Uhh, non," he said simply.

"Oh, okay," I said and sat back in my seat. A man of few words, this one.

Eventually, Pierre navigated the minivan up a darkened, narrow alleyway. The van struggled as we mounted a steep hill. After a few hundred meters, we stopped. Pierre put the van in park, yanked on the emergency brake, and hopped out. He pressed a button on the side of a large metal gate. After a few seconds, a man on the other side slid the gate open and Pierre motioned for me to get out of the van.

I half expected that Shawn would emerge, waiting to greet me. But it wasn't Shawn on the other side of the gate. It was the hotel's night watchman. It's understandable that Shawn's not awake, I told myself, at this hour. After I shook hands with Pierre and thanked him for the ride, the night watchman, who seemingly did not speak any English and very little French, helped me carry my luggage to my room.

La Maison du Pyla was everything Shawn had promised, even in its darkened state at one thirty in the morning. We entered through a set of double doors into the dining room/living room. There was a large oblong table on the left as we came in the door, and to my right I could see a couch and a small television. The varnished wooden floors were covered in a few striped, well-placed floor rugs and a small bookshelf held some worn paperback books and stacks of magazines. In an impressive show of strength, the night watchman hefted one of my duffel bags onto his shoulder and I followed him through a small hallway, dragging another duffel behind me. I would leave the last duffel downstairs for now—I didn't need anything from it tonight. We reached a set of stairs, and quietly—so as not to wake any other guests—climbed to the first floor.

The walls of my room were painted in coral reds and oranges and contained two beds—one single and one double. There was a desk pushed up against one of the walls and on it sat a large bottle of water. As the night watchman turned to go, I asked him, "Pardon, où est la toilette?"

He led me down the hall to where there were two small rooms—one that held a toilet and the other featuring a bathtub, shower, and sink. I would share these with other guests on this floor. I thanked the night watchman, made my way back to my room, and sat on the bed, exhausted. I could hardly believe that I'd made it. The emotions swept over me like a wave, and I'll admit it: I cried. I cried because I was exhausted. I cried because I missed Travis. I cried because I was nervous about what lay ahead. But most of all, I cried because I was overjoyed. I had made it halfway around the world from Toronto to Madagascar and my greater journey was about to begin.

Email received from Ken McGoogan
Re: Did you arrive safely?

Hey, Keriann: So I got up early, couldn't sleep: your dog, Cody's fault for scratching madly. Did a little work . . . while thinking about you on your way to Madagascar. I must insist, again, that you keep a journal. One trick: think of someone reading it on the other end. That's what Elisha Kent Kane did first time in the Arctic. He wrote his journal thinking of his brother, and then edited out most of the too-personal stuff later. Pop us an email when you can. I know you claim no news is good news, but as a former journalist, I am not so sure about that.

Love, Yer Dad

CHAPTER 4

Avenue of Independence

May 23, 2006

On my first morning in Tana, Shawn skipped breakfast. I had wandered downstairs around seven a.m., where I found a single place setting on the table alongside a spread of coffee, tea, juice, baguette, jam, and butter. Not exactly the hearty breakfast of stick-to-your-ribs steel-cut oats and fruit that I usually had at home, but the coffee was good, and that was what mattered to my jet-lagged system.

I heard footsteps approaching and a slight, middle-aged woman with short, curly hair emerged. It was Fanja, balancing a baby boy in her arms. "Welcome, Keriann," she said in an almost indecipherable accent, and introduced herself and her baby, Njacka ("en-za-ka"). I immediately felt comfortable with Fanja. She exuded kindness and sincerity. We chatted for a while as I sipped my coffee and I learned that Fanja, born in Tana, spoke fluent Malagasy, English, and French. She had lived in France for several years in her twenties. It was there that she had visited the Dune of Pilat, the tallest sand dune in Europe, and La Maison du Pyla's namesake.[1] I asked about her typical hotel patrons, and she told me that a lot of European couples in the process of adopting Malagasy children used to stay at Pyla, often for months at a time while they waited for their adoption process to complete.

"That's right," I said, remembering now. "Shawn mentioned that sometimes the place would be full of babies."

Fanja smiled, but then grew serious. "Yes, but the government has made stricter adoption laws, and so lately there haven't been as many *vazaha* adopting children. Now I see more researchers, like you."

As I swallowed the last bite of my baguette, I asked after Shawn, and Fanja directed me to his room on the second floor.

"You made it," Shawn got up from the desk after I had timidly knocked on his open hotel-room door. We shook hands. "Grab a seat." Shawn gestured to a second chair that was resting against the wall of his room.

"I've just been fine-tuning our research plan for Kasijy. Check these out." He pulled out a stack of laminated pages. "I had Andry create these lemur fact-sheets for our guides. See?" He held one up. "Each one has a photograph of the lemur species and a little information about it, written in Malagasy. These will help the local guides understand what we are looking for and will be a way for them to communicate which species they spot during our surveys. They can just point to the card."

In Kasijy Special Reserve, according to information published in 2001, we might see eight different species of lemur, four diurnal species and four nocturnal species.[2] The lemurs in Kasijy live *sympatrically*—a word I first learned way back in Brian Keating's class—meaning that they live in the same geographic location and so regularly encounter one another.[3] As I flipped through the laminated pages that Shawn handed me, I grew increasingly excited at the prospect of seeing more than one primate species on our journey, and possibly even more than one species at the same time.

I came across the cards for the nocturnal species first. There was the nocturnal fork-marked lemur. This species is known to be exceptionally vocal, particularly after dawn and dusk, which allows researchers to rely on their high-pitched, whistling calls to find and identify them in the field. Fork-marked lemurs also have a very unusual diet of gums, which has resulted in an especially long tongue and a specialized gut for digestion.[4] I hoped I would get to observe this species and their long tongues in action.

I flipped to the next card, which showed the gray mouse lemur. This lemur is among the smallest of all living primates, with a head–body length of just twelve to fourteen centimeters and a weight of fifty-eight to sixty-seven grams. Active at night, mouse lemurs are known to inhabit tree holes, and females tend to share their nests and form breeding groups to cooperatively raise their young.[5]

The next card showed Milne-Edwards's sportive lemur. I paused here for a chuckle at the comically large ears and huge, owl-like eyes in the picture. Like the mouse lemurs, these guys are also known to share sleeping sites, but when they forage, they go it alone.[6]

And the last nocturnal species was the fat-tailed dwarf lemur. These lemurs are notable because they go into torpor for six months or more during the dry season when food and water are scarce. During the wet season, they eat fruits that are high in sugar content to help them accumulate fat, which they store in their tails.[7]

"It's unlikely that we will see the fat-tailed lemurs," Shawn said when he saw me pause on the card. "We're in the middle of the dry season."

I nodded and flipped to the next card—a diurnal species! Staring back at me was the rufous brown lemur. A relatively large lemur—about two kilograms—this species is known for being sexually dichromatic, meaning that males and females are different colors.[8]

"You see," Shawn said pointing to the photos on the card, "the males are dark olive-gray, and the females are redder in color. It makes it easy to identify the group composition."

The next card was a real treat. It showed the rare eastern lesser bamboo lemur—quite possibly the cutest lemur I had ever laid eyes on. A delicate, olive-brown lemur with a round head and a short nose, its ears hidden by soft fur. With a specialized bamboo diet, this species also has a small geographic range, which is partly why it is listed as "vulnerable" on the IUCN Red List of Threatened Species.[9]

Finally, the cards I had been waiting for: two subspecies of Verreaux's sifaka. These were the lemurs I wanted to come back to study for my dissertation research. Von der Decken's (or Decken's) sifaka is a medium-size diurnal lemur, with a beautiful, pure-white coat. This particular species had yet to be studied in the wild because of its restricted range in highly fragmented deciduous forest in western Madagascar. I was keen to be the first researcher to collect behavioral data on this species. To date, all we knew was that Decken's sifaka is diurnal and lives in groups of about six to ten individuals.[10] There was also the possibility that we would see another member of the genus *Propithecus*, the crowned sifaka. Unlike the snowy-white Decken's sifaka, this lemur has a creamy-white coat with a chocolate-brown to black head, neck, and throat. Again, we had little information on this species, but we did know that it was classified as endangered due to forest loss.[11]

After I had perused the snazzy lemur fact sheets, Shawn and I talked about our plan for the day. We would grab a taxi downtown, Shawn said, so that we could go to the bank and cash our traveler's checks. While we were in town, he'd take me to the foremost French patisserie in Tana at the Hotel Colbert and out for lunch. Then, in the afternoon, we would go to the MICET offices, where I would meet our Malagasy student assistants and we could inquire after our research permits.

"We'll come back here for dinner," Shawn said. "Fanja hired a new chef. He's trained in fine French cuisine." His face lit up. "You should see the meals he's been preparing. They are practically works of art."

Taxis in Tana are beige-colored, and many of them are old Renaults. They don't have meters, so you'll need to agree to a price before getting in, and those prices are negotiable.

"Ready to go?" Shawn asked, entering the main foyer of La Maison du Pyla, where I sat waiting, his unofficial guidebook open in my lap. "Glad to see you are getting use out of that."

I stuffed the pages into my purse as Shawn led the way into the street. He wasn't feeling that well, he confessed as we walked, acknowledging something I had begun to suspect, given the amount of time he had been spending in his room.

"It's my gut. Plus, I have a bad case of Tana throat."

As the capital city of Madagascar, home to roughly 1.3 million people in 2006,[12] Tana is rightly known for having bad traffic and worse pollution. Shawn's throat issues, he said, were likely a result of the pollution.

"I have lots of throat lozenges if you need them," he offered.

But Tana is also a special city. In fact, as we walked out into the streets—my first time seeing the place in daylight—I was struck by its beauty. All around me, people were starting their day. Many looked as though they were off to work, dressed in skirts and suits, carrying briefcases. Schoolchildren dressed in powder-blue polyester uniforms meandered along the uneven sidewalks, spilling out onto the cracked concrete roads that surrounded the hotel. Some people sat outside storefronts, where slabs of meat were hanging, fully exposed to the elements. One man quietly strummed a guitar. I was engulfed by the sounds of traffic whizzing, car horns honking, children laughing, and bus drivers calling out schedules. I stopped for a moment and drank it in.

A king named Andrianjaka had built Tana in 1625 as a fortress along a rocky ridge. The city's hilltop views, looking out over the buildings and streets, can be quite striking. The name Antananarivo translates as "city of the thousand," and derives from the number of soldiers assigned to guard it during King Andrianjaka's reign. In 1793, the city became the capital of the Merina kings, whose huge stone palace, or Rova, still overlooks the city from the hilltops. A century later, in 1895, the French captured Tana and added infrastructure: roads, flights of stairs, gardens.[13]

Today, the city exists at three main levels: the lowest level, once ancient swamp, is today the downtown core; the intermediate cliff level, where the well-known Hotel Colbert is located; and the upper town, comprising palaces, cathedrals, and wealthy residential houses.[14] La Maison du Pyla is in Tsiadana, which nestles just below the intermediate cliff level. Tana is filled with winding cobbled streets and impressive staircases leading up and down between wonderful local markets. I was thrilled to be heading into the thick of it.

"There's a taxi," Shawn said, his voice drawing me out of my daze. He nodded toward a small cream-colored Renault parked on the opposite side of the street. We looked both ways and approached a car (my travel doctor back in Toronto would have been proud: "Travel hint number two: don't get hit"). I hung back near the sidewalk while Shawn stooped at the driver side door.

"*Hoatrinona Hotel Colbert?* How much for the Hotel Colbert?"

After some back-and-forth with the driver, Shawn gestured for me to hop in. I made to open the back passenger-side door closest to the curb, but the driver signaled for me to go around. He reached behind his seat and, using some force, opened the passenger door from the inside.

"The door handles are broken," Shawn explained as I got into the taxi and scooted over to make room for him. "The taxis here are, let's just say, interesting, and in varying levels of disrepair. Each one has a different problem—broken doors, windows, holes in the seats. A few years ago, I got in a taxi with a rusted-out floor. I could literally see the street whizzing past between my feet as we drove." He raised his eyebrows. "That ride was a bit unnerving"

Shawn settled into his seat, all six feet five of him. His neck was bent awkwardly forward, and yet his head still touched the roof of the taxi. He had his knees folded into his chest. Shawn saw me noticing and grinned: "These taxis are not built for guys like me."

As the taxi lurched through traffic, I looked out the window. It was a beautiful, sunny day and we were surrounded by hills covered in traditional red-earthed houses. Cars and motorcycles sped past, weaving in and out of their lanes. The rules of the road were unclear, but certainly the use of one's horn was widely accepted.

We drove through a large tunnel, and when we emerged passed a Hollywood-style sign that read "ANTANANARIVO" in large block letters. It was nestled on a cliff below what must have been the Rova. I snapped a few photographs with my pocket-sized Nikon Monarch digital camera that Travis had bought me. Suddenly, and without warning, the taxi slowed, and pulled into a gas station.

Shawn sighed and turned to me, "Don't worry, this almost always happens."

What always happens?

The driver stopped the car and looked back at us, expectantly. Shawn handed him one green and one purple paper bill—3,000 ariary, Madagascar's local currency, worth about $1 Canadian. The driver took the cash and reached under the front passenger seat, pulling out an empty one-liter plastic water bottle that still had its paper wrapper on it. "Eau Vive," it read. As the driver got out, Shawn explained that most taxis don't keep full tanks of gas, but gas-up once they have a fare.

A few minutes later, we were back on the road, though now we were stuck in gridlock. "Typical," Shawn muttered. We inched our way along, passing a small park situated in the middle of two lanes of traffic. Up ahead I could see restaurants, clothing boutiques, and electronics stores.

"This is the downtown core," Shawn said. "We're on the Avenue d'Independence, the main shopping drag. Back there," he pointed, "are the open-air markets. They're worth checking out—you can buy tons of great local crafts. Just be mindful of your belongings if you do go." Shawn held up his backpack. "The markets can be packed with people—a perfect venue for pickpockets. Someone actually cut a hole in the outside pocket of my pack a few days ago. Luckily, there was nothing in that pocket, but I had to sew it back up."

We were moving very slowly now, stopping and starting, as the traffic grew thicker. Outside our windows, various vendors walked the street, meandering from car to car; displaying their wares on oversized pieces of cardboard they held with two hands. I peeked at the goods, careful to avoid eye contact with the vendors—I didn't want to get their hopes up as I wasn't going to buy. The variety of available goods was astonishing. In the span of one city block, I could have purchased new sunglasses, fresh bathroom towels, a calculator, and a plastic bookshelf. Oh, and the vendors were very keen to sell us a slim Malagasy-English dictionary that looked as though it had been printed on someone's home computer.

Then something caught my eye that I wasn't prepared for. I watched as two young girls who couldn't have been older than six or seven, one with a baby strapped to her back, approached the vehicle in front of us, hands outstretched. The children were dirty, skinny, obviously malnourished. One girl's gray dress was literally in tatters and neither girl wore shoes. The baby had his or her eyes closed and head hanging limply backward.

Before leaving home, I had read that Madagascar is one of the poorest countries in the world. The World Bank reported that in 2006, 75 percent of Malagasy people were living below the national poverty level.[15] I had read that many children and their families live in extreme destitution and cannot meet even the most basic needs, such as access to safe drinking water, proper sanitation, access to medication, and sufficient food.[16] In that moment, sitting in the taxi on the Avenue d'Independence, I learned that there is a huge difference between knowing that such poverty exists and seeing it firsthand less than five meters away.

A rush of emotion washed over me, and even now, more than a dozen years later, I find it difficult to articulate what I was feeling. Certainly, I felt a deep sadness and sympathy for those children. Even if they had parents, they were so clearly malnourished that they must go to sleep every night with hunger pains, and they probably battled worse ailments. They spent their days wandering the streets of downtown Tana, begging, when they should have been laughing and playing with their friends. I thought of my own childhood, filled with happy memories—riding bikes with my friends, my mother teaching me how to bake cookies, my father reading to me from Roald Dahl as I drifted off to sleep. These children would never know such memories. We were different not because of anything I had done, but because I had chanced to be born in Canada to middle-class parents.

As I wrestled with these complex emotions, I wondered: Am I a responsible tourist if I give money to a child who appears to be suffering, or will my giving create a business of begging where one wouldn't normally exist? I was aware that many visitors before me must have struggled with this same question. Will this child die without my help? Would it be better to contribute at a larger scale? That day, in my heart, I knew that giving a few ariary to these three children wasn't going to solve anything. I did it anyway. And our taxi rolled onward.

Some years later, I made a decision that I could feel good about. My trusted Bradt travel guide—the bible of Madagascar guidebooks—had mentioned a charity run by the Sisters of the Good Shepherd in Tana.[17] The nuns run a school for children whose parents cannot afford to pay for registration, uniforms, and books. These are the children who, if left without assistance, take to the streets as beggars. Although I don't identify with any religion, I would visit that school to see what they were about. I met the children that the nuns teach each day and feed each morning. I met the women who have found employment with the sisters by creating beautiful cross-stitched fabrics, including napkins, tablecloths, and pillowcases. Now, every time I pass through Tana, I buy some of the cross-stitching and donate to that school. And every time I do, I think of those three children that I saw on the Avenue of Independence on my first day in Tana.

If you need to go to downtown Tana, it is worth a stop at the patisserie in the Hotel Colbert. The patisserie is part of the swanky hotel (which boasts an indoor pool and spa) and offers an impressive selection of fine French bread, baguettes, croissants, pastry, and ice cream.

It took me awhile to get my head right after seeing the three street kids, but I managed to shake it clear. Our first stop after the taxi dropped us off was the patisserie in the Hotel Colbert. Shawn said that we'd better fuel up before the long bank lineup that no doubt awaited us. I had a tough time deciding among the dozens of brightly colored French pastries on display, but in the end, fighting down thoughts of those children, I went for an éclair and paired it with a café au lait—classic.

After guzzling our coffees and scarfing down our pastries, we headed to the bank to cash our traveler's checks. We had to take out all the money we would need in the field because we wouldn't be stopping at any other banks along the way. Shawn led the way out of the patisserie, along the broken sidewalks, and down a large flight of concrete steps to the Bank of Africa building. The building was quite large, with rows of yellow taxis parked outside and groups of people loitering outside on the steps. As we made our way inside, Shawn explained how it would work.

"It's a really strange system here," he said. "You remembered to bring your traveler's checks and the receipts?"

"Yes," I said and pulled out my envelopes.

"Good. It's really not great to carry both the checks and the receipts, since the receipts are meant as backups for lost checks—I am sure your bank told you that. But, for some reason in Madagascar they won't accept the checks without the receipt. Go figure."

"Should we get in line?" I nodded my head in the direction of a group of people. "Wait . . . *is* that a line?"

Shawn smiled. "No, that's the other strange thing." He paused. "Hand me your passport."

I pulled my passport out of my travel purse and handed it to Shawn. He took it and said, "You don't stand in line here. Your documents do." He carried our passports to the front counter and placed them next to the other stacks of papers. When he returned, he said, "It might be awhile."

After about twenty minutes or so, the teller called us up to the front. Shawn guided me in requesting small bills—in rural areas, getting change for anything larger than 5,000 ariary is next to impossible. When all was said and done, we walked out of the bank with bags weighed down with stacks and stacks of local currency. At an exchange rate of about 2,700 ariary for each Canadian dollar,[18] and even though we had each cashed only $700 Canadian, we emerged from the bank millionaires.

Now I faced the problem of how to carry my embarrassing wads of money. I managed to put them as discreetly as I could into the bottom of my travel purse, which I now clung to for dear life. As we made our way back to the Tana streets, I could feel the money weighing down my bag, practically bursting a hole through it. I was nervous about navigating the streets with more money than the estimated per capita GDP in Madagascar (about $400 US),[19] especially after the pickpocket story Shawn had relayed in the taxi. And there was something else. A guilty feeling in the pit of my stomach. My mind flashed to the street kids we had seen earlier, begging for just a few ariary. Hungry. Wasted. This was by far the poorest country I had ever visited, and the poverty that I had seen in just that first morning was almost enough to bring me to my knees.

I pushed onward, and we spent the rest of the morning visiting an internet café and gathering more supplies. Shawn helped me find a telephone cord and converter jack to use for some good ol' dial-up internet at La Maison du Pyla, which was not yet wireless. Back at Pyla the night before, I had also had the anxiety-inducing realization that I had failed to pack a power plug adapter, rendering my laptop useless—a huge oversight, considering how long I had spent planning and preparing. Luckily, Shawn knew where I could buy one. He also helped me pick out a surge protector, warning me that Tana was prone to many a power outage, which had wreaked havoc on some of his electronics in the past. We hit up the local grocery store, ShopRite, to pick up some snacks and drinks for later. I don't know what I was expecting from a Madagascar grocery store, but the ShopRite had an impressive selection. I bought a cheap bottle of red South African wine, a fresh baguette, some cheese, and chocolate—the necessities.

For lunch, we stopped at one of Shawn's favorite restaurants in Tana, Chalet des Roses, a small bistro that served pizzas and other Italian fare. As the waiter seated us at our white-clothed table, I couldn't help but feel disappointed that my first meal out in Tana was Italian food. When I am traveling, I prefer to sample local cuisine whenever possible. As I took my first bite of pizza, I resolved not to tell Travis about this. He was the one who had converted me to local cuisine.

Before we started dating, after the Belize field school that we were both a part of, Travis and I and five other girls had spent twelve days traveling through Belize and Guatemala. We had made our way through San Ignacio and across the border into Guatemala to visit the Mayan ruins at Tikal. We'd stayed in Antigua and enjoyed venturing into several local markets. The other girls and I spent hours in the crafts markets, poking through the scores of colorful local fabrics and intricate wooden carvings. Travis took a different tack. He disappeared into the food market. He would emerge periodically, as if from nowhere, each time holding a bag filled with new, mysterious fruit, and chowing down on a different kind of street taco.

"Want to try some?" he'd always offer.

While some of the other girls cringed, I would usually partake, and in so doing had my eyes opened to a new kind of travel. I now firmly believe that it is imperative to sample the local cuisine when visiting a new place. The culture of food is powerful, and taking advantage of it provides a deep connection with the local people that you can't get any other way. To this day, it doesn't matter where Travis and I wind up, whether it's Kentucky, Montreal, South Africa, or Thailand, food always plays a major role in our itinerary.

I relayed my desire to sample local Malagasy food to Shawn as we sat sipping our Italian wine, and he assured me that there would be plenty of time for that.

"Once we hit the road," he said, "you will have all the Malagasy food you could ever want."

The MICET offices are in the Manakambahiny (man-a-kam-ba-hee-nee) district of Tana. On your first visit to the offices, be sure to introduce yourself to the wonderful staff and remember to pay your $300 facilitation fee. Our team also rents out a space in MICET's storage lockers, where we keep our field equipment.

Andry ("an-dree") and Sahoby ("sa-hoo-bee") were waiting for us in the MICET office. After lunch, Shawn and I had made our way there by taxi, and when we entered the sparkling clean office, two young men—they looked about my age, mid-twenties—were sitting at a large rectangular table near the front windows. Acting on Shawn's instructions, the two University of Antananarivo students had spent the morning purchasing and organizing our supplies for the field.

"Manahoana, Shawn." One of them smiled and stood to greet us.

I shook hands with the two, who were soon-to-be our traveling companions and field assistants. At this time, a committee of representatives from CAFF/ CORE was overseeing all research in Madagascar. CAFF/CORE was made up of representatives from Madagascar National Parks, the Ministry of the Environment, Forests and Tourism, and the Ministry of Higher Education. This committee required that every foreign researcher working in Madagascar should hire one student assistant from the University of Antananarivo.

In our case, each of us would support one student from the Paleontology Department. We would pay their school fees, provide them with field equipment, and help support them academically as they completed their "DEA," or Diplôme d'Etude Approfondie (the equivalent of a master's degree). The relationship was mutually beneficial, Shawn had said, and I would soon learn firsthand that Malagasy student support is invaluable. The students were experts at navigating local markets during supply runs. They acted as unofficial Malagasy–English translators. They helped with data collection in the field and, most importantly, they knew how to navigate local politics in a way that we never could.

At five feet seven, I towered over Andry and Sahoby, who both had the typical Malagasy build—short and slim. Standing there, even though I am medium-size in Canada, I felt like a giant heifer. Andry, the more outgoing and articulate of the two students, had worked with Shawn for the past five years. He had a friendly face and kind dark eyes. Shawn had trained Andry well during their many field seasons in the southeastern part of Madagascar and trusted him now to take the lead in the Kasijy preparations. The older and more reserved Sahoby was new to Shawn's team, and had not yet worked as a field assistant. Both men spoke fluent English, French, and Malagasy.

"Andry is a 'Tana kid,'" Shawn had told me in the taxi ride to MICET.

Shawn explained that Andry was born and raised in the capital city and came from an educated family of professionals. "His uncle's a doctor, and so Andry has had more opportunity than someone who was born in a rural community, like Sahoby. You'll probably notice a difference between the two—Andry is a bit of a city-type and doesn't always enjoy the forest. I don't know Sahoby that well yet."

"Benjamin wants to speak with you," Andry told us after our introductions. "I think he is worried about Kasijy."

Shawn sighed heavily and led the way up the corkscrew stairs to a row of offices.

Dr. Benjamin Andriamihaja, the director-general of MICET, stood up from his desk as Shawn rapped on the door and we four poked our heads into his office. He wore a gray suit, a yellow button-down shirt, a gray tie, and a smart pair of glasses. His handshake was no "limp fish" either; it was firm and powerful. Anybody could see that Benjamin was intelligent and charming, and held his director position at MICET for good reason.

Politely, he invited us to sit. His English was impeccable. He pronounced each word with precision and clarity. After our introductions and a bit of small talk, we got down to business. Benjamin and the team at MICET had been working tirelessly for the past two months to secure our research permits for Kasijy.

"Your permits are nearly ready," Benjamin said, "but the representatives from Madagascar National Parks are concerned about your safety."

He explained that because Kasijy was so remote, and because *vazaha* researchers had never explored the area, the park officials would require that a CAFF/CORE representative and four armed members of the Madagascar military, gendarmes, accompany our team.

"They are concerned that you will encounter bandits between Kandreho and Kasijy," Benjamin said, plainly. "To be honest, Shawn, I also have concerns."

That Shawn wasn't thrilled by this news was written all over his face. At lunch, he had told me that a few days ago, he had finalized the food and supply list for our crew, which Andry and Sahoby had already begun to organize. He had planned enough food for him and me, Andry and Sahoby, a cook, and three local field guides. We would also hire about ten porters for a few days during the hike to carry our gear to the field site, which he had factored into the budget and supply list.

Now, Benjamin was indicating that we would need to purchase supplies for five extra people. Not only that, we would have to reassess how many vehicles we needed. Through MICET, we had arranged for two four-by-four vehicles, together with drivers, to carry us to Kasijy. To transport the extra bodies and supplies—the CAFF/CORE representative would be coming with us from Tana, and we would pick up the gendarmes in a smaller city in the central-north, Maevatanana—we would need at least one more vehicle. That meant paying for the vehicle, driver, and fuel.

Shawn asked whether we would require the CAFF/CORE representative and the gendarmes for the entire time. "Maybe they could just accompany us for the first few days while we make our way to the field site and get set up. And don't you think four armed guards is overkill? I've worked in remote sites before. Couldn't we get away with two?"

Benjamin listened calmly and professionally, nodding and jotting down notes on a large yellow legal pad. "Okay," he said as we stood to leave. "I will look into all this and try to have your permits and vehicles ready tomorrow."

We spent another hour at MICET while Andry and Sahoby showed us the supplies they had purchased so far, and the itemized list they'd started. The MICET storeroom was filled with 140-kilogram plastic rice sacks—Shawn's field-luggage of choice ("The sacks are easier for the porters to carry, so long as we limit their weight to twenty-five kilograms each"). The sacks contained various field supplies including plastic buckets; dishes; cans of vegetables, meat, and fruit; and bars of Madagascar's famously delicious Robert's chocolate.

"I never go to the field without a hefty supply of Robert's," Shawn said when he saw my eyes widen at the box of fifty chocolate bars. "It helps keep the morale high."

Having tasted it, I understood. The Chocolaterie Robert was established in 1937 by a French couple from Reunion Island—Mr. and Mrs. Robert.[20] Robert's chocolate is made from Trinitario and Criollo cocoa beans, which are grown in the northwestern Sambirano region of Madagascar.[21] The Chocolaterie prides itself on its "bean-to-bar" model, processing cocoa into chocolate within days of its harvest.[22] The Madagascar chocolate we can purchase in North American grocery stores, on the other hand, is made from exported cocoa beans, which may sit for months before they are processed. The Chocolaterie provides employment for many Malagasy families and has recently started exporting small amounts of chocolate to Great Britain. In 2014, Robert's chocolate would win gold at the world finals for the International Chocolate Awards for its 50 percent cocoa bar.[23]

Now, before we hopped in our final taxi of the day back to La Maison du Pyla, we arranged that tomorrow a MICET driver would take us to the local food market, where we would stock up on arguably the most important of all the supplies we would need: rice and beans. Shawn turned to me, as he had done in his office back in Canada, and said it again, chuckling to himself: "Hope you like rice."

———————————————

Email received from Travis Steffens

Re: Re: Fine French cuisine

Hi honey,

Sadly the two pictures never made it . . . you might need to downsize them before you email. The high resolution on the camera makes for large file sizes. You packed in a lot yesterday! Tana sounds pretty awesome, even with the pollution. You should be used to that anyway—didn't you spend the past year in the "big smoke"? It sounds like your hotel's food is top notch. I know how you love crème brûlée. No wonder Shawn only wants to eat at the hotel for dinner.

Today was unproductive monkey-wise but I tried, and that's the point. I also talked with my brother and he is game for making sure the bar is taken care of for our reception in Calgary. Are you as excited as I am with our decision to go to Thailand for the wedding? I hear the street food there is amazing! I miss you a LOT!!!

Love, Travis

———————————————

CHAPTER 5

Big Guy Down

May 24, 2006

On my second day in Tana, Shawn skipped breakfast. As I sipped my corossol juice (the fruit tastes something like apple and cinnamon), Fanja came in to tell me that Shawn wasn't feeling well and wouldn't be coming down at all. "He said you should go with Andry and Sahoby to the market and help them organize the supplies at MICET. He said you should all go for lunch at the Indonesie. I can give you directions." Shawn had also asked that I check on the status of our research permits.

We were due to leave at nine a.m., so I had a couple of hours to kill at the hotel. I decided to try the internet. I checked with Fanja that the phone lines were free and used the phone cord that I had bought the day before to connect my computer to the phone jack on the wall in the dining room. It took a few tries, but finally I heard the old familiar crackling, hissing, and screeching sound of the modem connecting to the dial-up internet. I hadn't heard that in a while. For a moment, as I opened up my browser, I was transported to 1994. Junior high. I could use my family's home phone connection to a newfangled thing— the internet. I would spend hours in chat rooms that my brother had alerted me to, happily typing away, sometimes to *eight* other people. That's right, eight. Now, in Madagascar, the dial-up connection was slow, but I managed to send a few messages to friends and family back home.

Though slow internet connections can be frustrating—especially because we've all grown so used to instant responses—my experience in the field has made me grateful for any connection whatsoever. In Monkey River, for example, during my master's fieldwork, I had no access to the internet or telephone at our cabin. Every Sunday, I would kayak one kilometer across the river to Monkey River Town, where I would line up with the other locals to use the "village phone," a line that was shared by the entire village. When my turn came, I would pull out my calling card, punch in the required codes, and connect with my family for ten or fifteen minutes. As for internet, I would have to wait for the monthly visit to Placencia, where I would spend countless hours in internet cafés sifting through a month's worth of email messages. A few years after my field season, the site in Monkey River would become downright luxurious when it received a cell phone signal.

So, I thought the dial-up connection at La Maison du Pyla wasn't too bad. Once we headed for Kasijy, our only connection would be the satellite phone that Shawn had purchased, along with a phone plan that would allow us to send and receive short text messages and make phone calls in emergencies. We would not be making monthly visits to internet cafés. We would have no internet access until we got back to Tana. I reminded my parents and Travis of that in my emails.

I was shutting down my laptop when Andry and Sahoby arrived.

"Manahoana, Keriann," Sahoby said as he walked into the foyer.

"Non, *Salama*," Andry corrected, with a grin.

Manahoana is "hello" in the Merina Malagasy dialect, which is also the official Malagasy language dialect in Madagascar. *Salama*, on the other hand, is "hello" in the Sakalava dialect. Both Andry and Sahoby are Merina—born and raised in the highlands of Tana. Andry's correction was intended to prepare us for Kasijy, in the northwest, where the people came from the Sakalava culture.

According to my trusty Bradt travel guide, archaeologists believe that the first people arrived in Madagascar from Indonesia/Malaya about two thousand years ago.[1] Bradt's estimate for when humans arrived in Madagascar aligns with known evidence of butchery on the bones of extinct megafauna combined with the discovery of pollen spores of what is believed to be a human-introduced species. This evidence indicates that humans arrived just a few centuries before the beginning of the first millennium, around 2,300 BP (before the present).[2] However, there is some evidence that humans may have arrived to the island much earlier. At sites in the interior of Madagascar a research team discovered elephant bird specimens that showed evidence of butchery, and these bones were dated to ten thousand years ago.[3]

Looking to the genetic evidence, to date we cannot support such an early human colonization, though it is possible that these early human populations didn't leave a genetic signature.[4] The genetics can tell us more about where the early Malagasy populations came from, however. Researchers have found evidence of both Austronesian and Bantu descent, with a split from South Borneo between 3,000 and 2,000 BP, and a more recent genetic split from Southern African Bantus occurring about 1,500 BP.[5] After humans had made their way over to the island, there was a rapid expansion of the breeding population, beginning around 700 CE.[6] In addition, there was a growing trading network that connected Madagascar, China, India, Indonesia, Arabia, and Africa, leading to the spread of villages and hamlets, the introduction of new crops, and the keeping of livestock.[7]

Today, the island country is incredibly culturally diverse. The Malagasy people are said to be "progeny of the Indian Ocean," and historical, linguistic, genetic, and archaeological evidence combined point to origins in Southeast Asia, East Africa, South Asia, and the Near East.[8] Madagascar's rich cultural history has produced roughly eighteen different ethnic groups, each with its own traditions, beliefs, art, and language dialect.[9] The Merina and the Sakalava are the largest and best known.

Early on, Madagascar's diverse human population was organized into several small, ethnically divided communities that relied on trade, agriculture, and cattle herding.[10] But the sixteenth to eighteenth centuries saw higher populations and more centralized political systems, which resulted in a dynastic class of rulers.[11] This brought about *Maroserana*, a series of large-scale kingdoms in the southern and western regions of the island.[12] The Maroserana kings adopted and spread the cultural traditions of their subjects while expanding territorially.[13]

The Sakalava ethnic group became dominant in western Madagascar and is now the most widespread tribe in the country.[14] The Sakalava are known for royal ancestral worship and are thought to have first emerged in the region known as "Menabe" on the central west coast around 1660.[15] Eventually, they moved their capitals to the interior, but they continued to control the coastal ports where they ran military campaigns and exchanged captives for guns, gunpowder, and other commodities.[16]

Meanwhile, in the central highlands, the Merina kingdom formed with the union of several small chiefdoms.[17] This kingdom had the most stratified caste system in Africa.[18] The marshes in the central highlands enabled rice to become an important dietary staple.[19]

"We will go to the market first, for the rice and beans," Andry said as he led the way outside. The familiar blue MICET van was waiting, Pierre at the wheel.

Rice and beans are the two most important foods for fieldwork in Madagascar. If you don't provide rice, in particular, your Malagasy team will be very unhappy and may even quit. Rice is served at every meal. It is best to purchase your rice and beans in Tana from the Analakely market.

"Watch your belongings," Andry warned me. "There are lots of pickpockets."

We climbed out of the MICET van at the Analakely market, the largest in town. Known also as the Zoma ("zoo-ma") market, it comprises a series of clay-roofed pavilions that shelter numerous stalls offering an eclectic mix of products including clothing, sunglasses, fruits and vegetables, rice, fish, and meat. As I stepped out of the van, I felt overwhelmed by sights, sounds, and smells. The market teemed with vendors and shoppers. All around me, people were animatedly speaking in Malagasy, laughing, and haggling. I followed Andry and Sahoby deeper into the market, where we made our way to the food stalls. The vendors displayed their wares in baskets and barrels—colorful fruit, vegetables, beans, and rice. Andry led the way toward one of the vendors, a gray-haired Malagasy woman who sat perched atop a stool behind several barrels full of beans. The smell of the meat and fish wafted over us from a few aisles away.

"What kind of beans do you want?" Andry asked. "Shawn said we need 250 kilograms total."

I'll admit the question caught me off guard. I had given no thought to the *kind* of bean I would prefer. In Belize, there hadn't been much choice. The typical Belizean bean type was a standard kidney bean. Belizeans did vary the way in which the beans were prepared, however, and we graduate students had a running joke that you were either a "rice-and-beans" person or a "beans-and-rice" person. To make "rice-and-beans," you cook white rice and kidney beans together with onions, garlic, spices, and oil in one pot. "Beans-and-rice," on the other hand, involves stewing the kidney beans slowly with onions, garlic, spices, and oil until they create their own sauce, which is then ladled on top of the rice. For the record, I am a rice-and-beans person. Travis is a beans-and-rice person. It is a point of contention in our marriage.

I surveyed the different beans on display at the Madagascar market. There were, of course, the beans that I knew, the safe choice: red kidney beans. There were also white kidney beans, green lentils, black lentils, chickpeas, broad beans, and black beans. I could feel the sweat pooling on my forehead. Which bean to choose? The pressure!

"We can get five different kinds," I suggested. "Fifty kilos of each? Is there any kind of bean you don't like?"

After some discussion, we decided on fifty kilograms each of red kidney beans, white kidney beans, green lentils, chickpeas, and broad beans. As for rice, the only decision required was which vendor to choose, though there were many. Andry took the lead, carefully combing through the barrels on display to find the one that was the best quality, with the fewest rocks.

Rice is culturally significant in Madagascar. Besides Asia, this country has the longest history of rice cultivation.[20] Put simply, rice is the staple food of Madagascar. Though the earliest inhabitants did not leave a written record, researchers have used linguistic records and an understanding of monsoon wind direction to deduce that crews from Southern Borneo were the first settlers to the island.[21] And they brought rice and rice culture, as far back as 400–1000 CE.[22]

Data from 2002 indicated that rice made up 48 percent of Madagascar's total calorie consumption and amounted to a per capita consumption of about 95 kilograms per year.[23] That means, a family of four would go through about a kilogram of rice each day. The Malagasy also use rice in ceremonies, and give it to visitors, workers, and others passing through. On the outskirts of Tana, I had already seen large swaths of irrigated rice fields. From above, I learned later, these look like a checkerboard in every hue of green imaginable.

Although beautiful, Madagascar's rice fields are a concern for conservationists, especially where lemurs and other endemic biodiversity are concerned. Slash-and-burn agriculture, or *tavy*, is the traditional system here. With *tavy*, primary forest or secondary vegetation is cut and burned, and upland rice is cultivated for one season, followed by a root crop such as manioc or sweet potato. From a human perspective, *tavy* is beneficial because the nutrient-rich forest soils produce a high crop yield in the first year. Also, the system is affordable and allows farmers to produce a variety of different crops (not just rice).[24] Given its long history in the country, *tavy* is deeply rooted in Malagasy culture and tradition.

After that first good year, however, soil fertility declines rapidly, and the land is left to become fallow. Frequent, uncontrolled fire can also become problematic because the system kills native tree species, allowing for the invasion of exotic, shrubby plant species. Repeated use of *tavy* produces vast areas of grassland that are subject to erosion, no longer viable for agriculture, and uninhabitable to most lemur species. While *tavy* is sustainable in countries with low human population density and abundant land, it is problematic in places like Madagascar.[25]

According to the World Bank, the 2006 population density of Madagascar was 32.4 people per square kilometer, while in that same year, Canada's population density was a measly 3.58 people per square kilometer.[26] Today, *tavy* is recognized as the main cause of deforestation in Madagascar,[27] and as a huge threat to the survival of forest-dwelling lemur species.[28] The problem becomes even more complex when you consider the needs of the human population, 80 percent of whom depend on agriculture to survive.[29] And so, at the market, as the vendor carefully measured out our 250 kilograms of rice, I watched with mixed emotions.

Our rice order filled, we visited the stall two doors down and bought forty kilograms of carrots, thirty kilograms of cassava, fifteen kilograms of garlic, and fifteen kilograms of onions. We hired four porters to carry our purchases to the MICET van, where Pierre was waiting. The van practically groaned when the porter tossed the final heavy bag of rice into the back. This was serious business. It hit me then that to warrant that supply of food, Kasijy must be even more remote than I had imagined. In Belize, we would only ever have to think about food for two or three people over the span of one month. And if we were ever in a pinch, we could find a small selection of canned goods and vegetables from a local shop in nearby Monkey River Town. Plus, once a week a vegetable truck would drive by, allowing us to stock up. What was I getting into here? I shook it off and hopped into the back of the van. Next stop: Jumbo Score.

A Walmart-style superstore on the outskirts of Tana, Jumbo Score contrasted sharply with the Analakely market. We entered the gated parking lot, which seemed empty and sterile. We hopped out of the van and crossed the paved lot. We had already bought most of the necessary supplies, but because we were bringing gendarmes, we needed to pick up a few more cans.

Inside the store, we encountered row upon row of neatly stacked consumer goods—everything from electronics to clothing to food. We made our way into the canned goods aisle where Andry read from his neatly handwritten list, checking off items as Sahoby and I filled several large boxes with cans of sardines, corned beef, jam, and milk powder.

"Whoa," I said, and abruptly stopped packing. Suddenly, the room was spinning.

"Are you alright?" Andry asked, a look of concern on his face.

I squatted down on the floor. "Just a little dizzy," I said, and put my head down between my knees.

I had felt this feeling before. As a child, I had been prone to fainting spells. The first incident I can recall was when I was in Brownies. I was about seven or eight and my Brownie troop was learning all about first aid. The troop leaders had brought in some presenters from St. John's Ambulance, who were demonstrating what to do in emergency situations. I don't remember what the precise discussion was, but I do remember very distinctly the mention of blood. There may also have been some red props, although that could have been my mind running away. Regardless, I fainted.

"That's when I knew you probably weren't going to be a doctor," my mom used to tell me. Don't worry, Mom—it turns out, there is more than one kind of doctor!

The dizzy spells continued. When I was ten, I hit the floor again after basketball practice one day when I was wrapping a small cut on my hand. As a teenager, I would often get dizzy and clammy when I was on my period. I worked at the Calgary Stampede one summer as an usher for their famous Grand Stand show. One hot day, after standing all day long directing the hordes of people to their seats, I hit the floor once more.

The problem would come and go. But it never persisted, so I just kept plugging along. It was Travis who eventually convinced me to see a doctor about my "spells." One evening, we visited our good friends, Katie and Nadim, in their southwest Calgary condo. We spent the night watching movies and drinking red wine. I had a few glasses but was by no means drunk—tipsy, maybe. At the end of the evening, while we said our goodnights at the door, and before I could get out the words "I don't feel so good," everything went dark. I came to on the floor of their apartment.

Fortunately, Nadim—an emergency-room doctor—had noticed something even before I did. Travis told me later that Nadim had calmly stepped toward me and gently eased me down to the floor as I lost consciousness. I was fine, but this dramatic incident was a tipping point. I made an appointment to see the doctor next day. After I described my symptoms, the doctor took my blood pressure and had me lie down and sit up abruptly a few times. He confirmed that, though it was in the normal range, my blood pressure was on the low end. Alcohol, he said, would exacerbate my already low blood pressure, and so it was no surprise that I'd fainted after a night of drinking. As he walked me out of his office the doctor said, "Just increase the amount of salt in your diet. It won't take much—just try and have a bowl of soup every day, for example."

"Lucky you," a nurse smiled at me as she breezed past us in the hallway. "You don't hear doctors prescribe high-salt diets very often."

Since then, I had learned to recognize the signs of a dip in blood pressure, and I would self-correct by chugging a Gatorade. This tactic was harder to do when I wasn't at home. Now, in the Antananarivo Jumbo Score, I chalked up the spell of dizziness to a lack of salt and made a mental note to add salt to my next meal.

After a few minutes, I stood back up. Right as rain.

To ensure that porters can transport your food supplies in the field, the packed weight of each bag should not exceed twenty-five kilograms. You'll be able to pack and store all of your supplies at MICET. I recommend that you number each bag sequentially and make a detailed inventory of what's in each. It's tedious but will ensure that you can quickly find what you need when you get to the field.

Back at MICET, we spent what was left of the morning organizing the gear into rice bags, each weighing twenty-five kilograms. We created itemized itineraries for each bag, numbering the bags, and tying them closed with twine. The work, which involved opening the sacks of rice and beans and redistributing the weight into new bags, was, as Shawn had anticipated in his unofficial guide, tedious.

I was still feeling odd, and twice more I had to sit down in response to a rush of dizziness.

"Maybe we should go eat," Andry offered. "Then Sahoby and I can come back to finish this work."

"Thanks," I said, relieved. "Shawn said he would treat us to Indonesian food."

Another one of Shawn's favorite Tana restaurants, located near MICET, was the Indonesian ("Try the nasy goring," he had said). The restaurant was empty except for the three of us. I got us a one-liter Coca-Cola to share, and we ordered our food. Eating did make me feel a little better. We talked about our past and upcoming adventures, and I grew more and more comfortable with these two kind university students. A feeling of relief washed over me, and I relaxed a little in my chair. We dug into our Indonesian rice bowls and Sahoby and I laughed together as Andry—who I now realized was a bit of jokester with a quick wit—relayed his previous experience in the field on one of Shawn's previous projects. He talked about how he had helped hire the field guides and coordinated the porters. Sahoby, more serious, spoke of the research project he wanted to do for his master's.

"Can you show me how to do behavioral samples?" he asked me.

"Of course!" I said and told them both about my prior experience with howler monkeys in Belize.

I knew now that the three of us students—although we were from opposite sides of the world—were in this together. In just a few short days, we would be on our way to the Madagascar wilds—together, as a team.

When the waiter poured the last of the Coke into our glasses, Andry offered a toast. "To Kasijy," he said, a look of sincere excitement on his face.

"Kasijy," Sahoby and I echoed, and we all clinked glasses.

That evening, Shawn came down to dinner. He was feeling a bit better, and I filled him in on the day's events. "The supplies are all packed and ready to roll. But I checked and MICET still hasn't received our final permits. Benjamin thinks they will arrive tomorrow."

"These permits are really late." Shawn looked concerned. "We should have received them by now. Malagasy people are really wonderful, but they don't like to deliver bad news."

"Is it cold in here?" I interrupted, reaching for my sweater, which was draped on the back of my chair.

"I'm okay," Shawn replied, and then continued. "We'll keep hearing that the permits are coming 'tomorrow'—and that's possible. But it's also possible, maybe even probable, that our permits may actually be weeks away."

We were scheduled to spend one more day in Tana. Shawn and I would pack up our personal items and purchase any last-minute necessities. So long as our permits did arrive, MICET would pick us up early from La Maison du Pyla on May 26 and we would hit the road.

Fanja emerged from the kitchen with our dinner—fish with rice. I sat for a moment, the plate of food in front of me. I picked up my fork. I put my fork down. Suddenly, there it was again—the dizziness. But this time, it was more than just dizziness. I felt wretched.

"I don't feel so good."

Part III

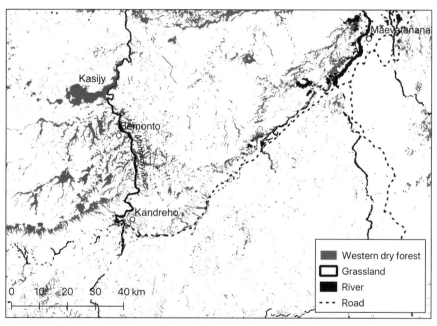

Maevatanana to Kasijy. *Map created by Travis Steffens.*
GIS data from J. Moat and P. Smith, Atlas of the Vegetation of
Madagascar / Atlas de la Végétation de Madagascar
(Richmond, UK: Kew Publishing, Royal Botanical Gardens, Kew, 2007).

CHAPTER 6

Slowly, Slowly

May 25, 2006

We would spend five more days in Tana, while I recovered from the worst food poisoning of my life. The fine French cuisine that Shawn and I had eaten at La Maison du Pyla—and later raved about to our friends and families—had made us both deathly ill. The chef had not been sanitary, and poor Fanja was mortified that both Shawn and I had gotten sick from the food.

In my Madagascar travel guide, Bradt explains that "most traveler's diarrhea comes from inadequately cooked, or reheated, contaminated food."[1] In fact, Bradt says, "Sizzling hot street food is likely to be far safer than the food offered in buffets in expensive hotels, however gourmet the latter may look."[2]

"I wasn't sure that it was the food until you got sick," Shawn told me later. "I didn't say anything because I didn't want to get the chef fired . . . in case it wasn't the food."

Fanja, a smart and responsible business owner, did fire the chef. And she was an amazing caregiver while I was laid up in bed—second only to my own mother. She checked on me regularly, bringing chicken soup, rice, and tea to my room. I spent day one of my illness running back and forth down the hall to the toilet, cursing the fact that my room didn't have a private bathroom. The following day, Fanja arrived at my door with a doctor.

"Keriann," she said gently, "Shawn asked me to call a doctor for you."

The doctor, a small-framed Malagasy man, wearing a suit and wire glasses, entered and I sat up in my bed. As he readied his kit, pulling out a stethoscope and thermometer, it became clear that he didn't speak English. He turned to me and as best I could tell, asked what was wrong. Realizing that I didn't know how to communicate what ailed me, I looked back up at the doorway in hopes that Fanja had stuck around, but she had gone. I sighed as I came to the realization that I would have to use my broken French along with a lot of pointing and gesturing to my stomach to explain.

I felt terrible. Not just physically, but emotionally. I was frustrated at this unanticipated communication challenge. Although I had aced my French exam back in Toronto and had diligently gone to Saturday morning refresher classes at the Alliance Française for nearly a year, I couldn't find the words I needed to explain what was wrong. I was struggling to remain composed. The doctor must have noticed. He gestured for me to hang on, walked over to his medical kit, and pulled out two different packets of pills and a few mysterious packets of powder. As he handed me the packets, he patiently and slowly explained—in French—what they were for. As best I could tell, the pills included an antibiotic and something for fever, and the powder packs were for nausea. All the meds were labeled, so I figured I could log on to the internet to double-check their purpose after the doctor left to ensure I had correctly understood.

Pills in hand, I thanked the doctor and retreated back to my bed, still in the throes of a pretty bad fever. Any primate researcher will confirm that getting sick during fieldwork is no fun. It's uncomfortable, sad, and can be extremely frustrating, especially if it gets in the way of data collection. Any primate researcher will also confirm that getting sick during fieldwork is inevitable. Taking ill in the first week of my trip to Madagascar was a new record for me, but the experience wasn't altogether unfamiliar.

Belize. The home stretch of my master's research. I had just two more weeks left of field time. There had recently been a flurry of excitement at the field site. Seventeen fresh-faced undergraduate students from the University of Calgary had descended upon us as part of Mary's annual field school. For the past ten days, Travis and I had hit pause on my data collection to work as teaching assistants, taking groups of students out to find the monkey groups and to practice behavioral sampling techniques. Greg Bridgett, another master's student and a close friend of ours, had also joined us. He and I would overlap for two weeks while I finished my data collection and he kicked off his. Meanwhile, Travis would depart with the field school students so that he could get to his job at the Plain-of-Six-Glaciers teahouse in Lake Louise.

A few days before Travis's scheduled departure, the three of us were sitting around the plastic foldout table in our cabin, about to take in a screening of *The Butterfly Effect* (note here that standards for what constitutes a good movie are significantly lower after six months in the field). About twenty minutes into the film, I started to feel feverish and had to excuse myself. I managed to sleep fitfully, but when I awoke the next morning, Travis took one look at me and furrowed his brow. "Your face," he said. "It's all swollen and puffy. Go look in the mirror."

I got up and hurried into the bathroom where a mirror hung above the sink. Even though I had been warned, I wasn't prepared for the image that stared back at me now. Travis was right. The entire right side of my face was swollen.

I looked like a lopsided chipmunk. I came out of the bathroom. "My mouth hurts," I told Travis. "I wonder if it's my wisdom teeth."

"Let's hop a ride to the doctor tomorrow," Travis said. "Meanwhile, you had better go lie down."

The next morning, the three of us headed to a small town called Independence where we knew of a private doctor, Pedro. He was a short, cute man, from Spain originally. One of our friends in Monkey River had recommended him highly.

"Open wide," Pedro instructed, as I sat in his examination room. He peered in at my teeth. As I sat there with my mouth wide open, Pedro explained to Travis that he had trained as both a doctor and a dentist. A stroke of luck!

Pedro instructed me to close my mouth, and as he removed his plastic gloves and threw them into the bin, in a thick Spanish accent he said: "The problem is seeemple. You have too many teeth," he paused. "And not enough room. You must go back to Canada and take out some teeth."

How could I ignore such plain advice? I changed my ticket and headed home earlier than anticipated to have my impacted wisdom teeth removed.

Now, at La Maison du Pyla, Fanja paid me another visit.

"Shawn asked me to give this to you," she said and held out a piece of paper. Shawn had very considerately spent some time on the internet deciphering the medications that the doctor had left with me. I sat in my pajamas on my bed and read through the list. First there was the antibiotic—turned out it was just Cipro. Funny, I thought. If I had known that, I would have just started taking the pills that I already had stowed away in my medical kit. Next on the list—the mysterious powder. Beside the medical jargon, Shawn had just scrawled: "Basically the same as Imodium." *Fantastic*, I thought. *I would love to stop running back and forth to the toilet.* I grabbed a glass, filled it with bottled water and dumped in one of the powder packets. As soon as I did, the smell hit me. How could something that was supposed to be for nausea smell so horrid? I pinched my nose and chugged.

"Ugh!" I couldn't help but exclaim.

That taste . . . was. . . awful. I've never tasted gym socks, but I am pretty sure that they taste a lot like that powder. I glanced around for something, anything, to get rid of that taste. Then I spotted the shiny black plastic packet poking out of my suitcase. My Sour Cherry Blasters. I had been saving them for when I was in Kasijy, but what the heck. I had nothing else available. I opened the pack and popped a couple of blasters in my mouth. Sweet (or rather, sour) relief. Opening that packet was the beginning of the end. Over the next few days, every time I took a pill, I popped a blaster. By Sunday, May 28, two full days after we were supposed to begin our journey to Kasijy, the packet was empty. So much for creature comforts while in the field.

But my lack of self-control with the cherry blasters was the least of my concerns. I was starting to feel better, but unfortunately it was looking like we would be delayed even further. Shawn—who now admitted that he had been sick since before I arrived—was still laid up in bed, emerging only sporadically to get a cup of tea or a slice of bread. He explained that he didn't like taking pills, and so while I was now on day two of my antibiotics and fully on the mend, Shawn was still attempting to battle the illness the old-fashioned way: through grit and determination.

But the truth was that we were running out of time. Every day that we waited for Shawn to recover was another day that he wouldn't spend in the field. Now that I was feeling better, I was getting bored and restless in Tana, and more than a little annoyed that Shawn wouldn't just go on a course of Cipro. The drug had worked so well for me and I honestly couldn't fathom why he chose to suffer. I wanted to get out there and see some lemurs. I think Tom Petty and the Heartbreakers sang something about waiting. That song was about lemurs . . . right?

I was struggling to keep my frustration at bay. The stakes were high. I had spent a lot of money to get to Madagascar. I had moved away from Travis, with whom I hadn't had contact for several days now. And the plan for the field felt like it was falling apart at the seams. So, when I saw Shawn on the evening of the 28th and he told me, "If I'm not better by tomorrow I will go on a course of Cipro," I was indescribably relieved.

Our revised plan was to spend Monday packing our personal belongings and doing our final chores in Tana. Andry and Sahoby would spend Monday with the MICET drivers, prepping the trucks and loading our food and supplies. Then, at seven a.m. on Tuesday, May 30, we would set out on the first leg of our journey. We still didn't have our permits in hand, but MICET had assured us that the driver would bring them when he picked us up. "I've told MICET that we need to be ready to roll when they get here," Shawn said.

The trucks were also to pick up Prospère, the representative from CAFF/CORE who would be joining us. Andry and Sahoby, who lived nearby at the university residences, would meet us at our hotel. We would then make our way to Maevatanana, where we would collect the two gendarmes who would come along for our final hike into Kasijy. With Benjamin's help, Shawn had convinced the authorities that Prospère and the gendarmes would join us only for the last leg of our journey. They would leave Kasijy with Shawn, thus allowing us to save money on food and supplies for the rest of our field season.

So it began. Andry and Sahoby arrived just as Shawn and I were finishing our breakfast. This was it: 7:15 a.m. on May 30, 2006. Andry announced that the trucks were waiting in the street. We were finally beginning our journey. I felt such a rush of excitement that I practically jumped up from the table.

Outside, Shawn's face fell when he spotted the familiar blue MICET van. "What's that doing here?" he asked, struggling to hide his frustration.

Andry explained that MICET had only two four-by-fours available today, so the third vehicle had to be the van. Shawn rubbed his temples, "This isn't going to work," he said. "That van can't navigate the terrain all the way to Kasijy. We need four-by-fours."

Andry shrugged and glanced at me, sheepishly. He looked back at Shawn. "I don't think we have a choice," he said. It was the van or nothing.

As the drivers added our personal items to the piles of rice sacks that were already strapped to the roof racks on the four-by-fours, Shawn looked around and then turned again to Andry. "Hang on. Where's Prospère?"

The drivers had neglected to pick up Prospère. Now, instead of hitting the highway, we had to navigate downtown Tana during morning rush hour to collect him from his house.

"Please tell me that the drivers brought the permits, at least?" Shawn asked Andry, exasperated.

No such luck. The permits were finalized, apparently, but still at MICET. After we picked up Prospère, we would have to head to MICET for the permits.

"Mora mora." Shawn sighed, throwing up his hands in defeat, and hopped into the front of one of the four-by-fours. "Mora mora," which means "slowly, slowly" in Malagasy, would become the mantra of our journey to Kasijy.

You are responsible for paying for all costs associated with the vehicles on the way to and from the field. Fuel can cost up to US$100 per day per vehicle.

We had done it. We had made it to . . . the gas station on the outskirts of Tana. Sahoby stood outside and was coordinating fuel and payments for each of the vehicles. I sat in the backseat of one of the four-by-fours with the shy Prospère to my left. Shawn looked back at me from the front seat.

"Check out Sahoby's scarf," he said with a grin. "Doesn't he remind you of a World War Two pilot?"

I laughed. It was true. Sahoby had paired the scarf with a puffy, faux-leather jacket. All he needed was a pair of goggles. Sahoby's silk scarf was gorgeous, though, and I would later learn that it was culturally significant. Silk production in Madagascar has a long history. In fact, Madagascar is the only African country that has a traditional silk sector.[3] There are two main types of Malagasy silk. The first derives from breeding silkworms of the species *Bombyx mori*, which were introduced in the 1800s, and the second is a wild silk from a native species, primarily varieties of *Borocera*, common in Madagascar's highlands and on the coast.[4]

Silk is produced mainly by women using small hand looms.[5] The process is intricate and involves a whopping thirteen steps after the silk has been taken from cocoons.[6] The women cook and wash the silk, pull apart the thread, wrap the thread around bamboo poles, dye the silk with natural colors, spin it onto a spool, and finally weave the final product.[7] Silk plays an integral role in Madagascar's local economy, and the silk sector is a source of income for many rural families, especially for women.[8]

And bonus! Silk production contributes to the protection of the environment because it provides a financial incentive for people to protect Tapia forests, home to the wild species of silkworm.[9] In fact, there are now organizations that work with rural farmers in Madagascar to help develop sustainable silk farms that serve to support both people and the surrounding ecosystem, such as SEPALI Madagascar and The Natural Silk Project.[10]

"Oh, I almost forgot," Shawn said and started digging into his backpack. "We should each take a spoonful of this."

He pulled out a plastic jar of powdered Metamucil.

"What for?" I asked. After just getting over my illness, the last thing I wanted to do was take something that would make me have to use the toilet more.

"On these long road trips, it's easy to get . . . gummed up. You don't have to, but I would recommend it."

I sighed and nodded. Shawn has done this before. He downed a spoonful of powder and took a good gulp of his bottle of water. He refilled the spoon and handed it to me. I did the same.

"Ugh!" I exclaimed after I had taken a sip.

Shawn laughed as I handed back the spoon. "C'mon, it's not *that* bad."

If only I had more cherry blasters, I thought.

Congratulations, you are ready to hit the road! Be sure to bring along a small flashlight, hand sanitizing lotion, and anything else you like to take on long road trips. I'd recommend bringing anti-nausea medication because the roads—even the highways—in Madagascar are long and winding.

Our three-vehicle convoy wound its way along Route Nationale 6, the primary paved highway in Madagascar. As we navigated the turns and periodic switchbacks, I watched the small concrete red-and-white kilometer markers that peppered the roadside. Our first stop would be Maevatanana, a town of about twenty thousand people located 150 kilometers north of Tana. Historically, the place is famous for its gold mines, which boomed in the nineteenth century. Shawn estimated that we would need about six hours to get there from Tana.

"Six hours?" I had exclaimed. "Six hours to go 150 kilometers? That doesn't seem right." I ran the math in my head to make sure I wasn't going crazy. A six-hour drive along a highway back in Canada would mean traveling at least six hundred kilometers.

Shawn smirked a little. He knew what I was thinking. "I know it seems odd," he said. "But these roads aren't like highways in Canada. The roads we are driving today are paved, but you'll see—they are very long and winding."

Having finally left Tana at nine a.m., with all team members and permits accounted for, we hoped to arrive in Maevatanana by about three or four p.m. When we arrived, we would meet with the local government officials and pick up our two gendarme escorts. It is considered unwise to drive Madagascar's highways at night, and so with the help of MICET, we had arranged to stay overnight at the home of one of the government officials. We had been told he had a large property where we could pitch our tents; it was in a nearby town called Mahazoma—a short ferry ride from Maevatanana.

As we got going, I learned that Shawn wasn't kidding about winding roads. The drive lulled my fellow passengers to sleep. Beside me, Prospère rested his head against the passenger window. Up front, Shawn wore his sunglasses and stayed plugged in to his portable CD player, his head bowed down to his chest. Me? Well, I couldn't sleep. I am prone to motion sickness. I could feel every twist and turn inside me. Each time the vehicle lurched—which was frequently—my stomach did a flip. I had been looking out of the window, counting the distance markers as we drove, but now that distraction was only compounding my nausea. I tried to focus my eyes on the horizon. As I did, I wondered, is it possible that I would see a lemur as we drove?

Back in Belize, we would sometimes spot howler monkeys as we made our way to town along the dirt road that ran adjacent to the forest. I closed my eyes and took a deep breath. I could see the lemurs now. I imagined a group of crowned sifakas, their snow-white bodies contrasting starkly against a lush, green tree canopy. This diurnal endangered lemur species is poorly known, and the largest populations survive in fragmented, dry deciduous forests between the Betsiboka and Mahavavy Rivers in northwest Madagascar.[11] I felt a rush just thinking about it. In Kasijy, we would camp along the banks of the Mahavavy. Group sizes of crowned sifakas can be up to eight individuals and the research team that had visited Kasijy before us had witnessed four individual crowned sifakas traveling alongside three individual Decken's sifaka.[12] Shawn had told me that proximity between species could mean that there was interbreeding, which could have implications for the taxonomic classification of crowned sifakas. Was there hybridization? There was so much to discover about this elusive primate. In fact, there were no data on the densities or habitat selection of any of the lemurs in Kasijy. Our study would help us learn more about the Kasijy lemurs—their behavior, ecology, and demography—and could be used to help us contribute to conservation strategies.

I opened my eyes. It only took a moment of watching the landscape carefully for me to realize that a lemur sighting wasn't going to happen. In fact, the landscape was not at all what I had expected. On the outskirts of Tana, I saw nothing but rice fields. These are beautiful in their vibrant green color, but they also represent ecological devastation. As we drove farther north, the rice fields petered out, but the forest that I had expected did not present itself. Instead, I saw rolling hills of sand and rock, sprinkled with shrubs. We would know we were near a village or a town when we saw a patch of tall trees, among them mangoes or eucalyptus.

Years before, during one of my shifts at the Calgary Zoo, where I worked for a summer drumming up funds for a new exhibit called Destination Africa, I had an encounter with a middle-aged woman. Now it came rushing back. I had been walking her through the new exhibit and pointing out features on the scale model that was housed in a portable building next to the construction zone.

"Here's where our rainforest exhibit will be," I said, pointing to the miniature trees under the glass case.

"Rainforest?" she scoffed. "I was in Africa a few months ago and there is no forest."

"Well, Africa is a big place," I said, trying to remain diplomatic.

After she left—without donating—I rushed over to my coworker. "Did you hear that?" I whispered, laughing. "No rainforest in Africa? She must have seen only the African savannah or something. I mean, think about it. Madagascar is practically all forest!"

Now, as we rolled through a stark landscape, my words came back to me. Clearly that woman had been out to lunch, but I had been wrong too. Madagascar is nothing like all forest, and the extent of habitat loss began to register.

I knew the reality, of course. Based on analyses of aerial photographs and Landsat images, researchers have found that between the 1950s and 2000, Madagascar lost 40 percent of its forest cover, and saw core forest reduced by almost 80 percent.[13] What forest remains is highly fragmented and inadequately protected. That level of habitat loss is one of two main reasons why Madagascar is considered a biodiversity "hotspot."[14] The other is that lemurs and much of the other amazing biodiversity that the island has to offer are found only in Madagascar and nowhere else in the world.

Identifying hotspots provides a way to set conservation priorities and enables conservationists to make the best use of the limited available funding.[15] The aim is to help the most species at the least cost—to get the most "bang for your buck." There are thirty-six areas that scientists have categorized as hotspots based on two criteria: they have at least 1,500 endemic vascular plants and 30 percent or less of the original natural vegetation.[16] As Conservation International puts it, a hotspot is both irreplaceable and threatened.[17] Sadly, Madagascar qualifies.

Much of the habitat loss that Madagascar has experienced is due to human actions, among them slash-and-burn agriculture, mining, and illegal logging of rosewood and ebony.[18] The problem is even more complicated when you consider that Madagascar is also one of the poorest countries in the world. The World Bank estimates that 72 percent of the people here survive on less than two dollars a day.[19] People are living hand to mouth and struggling to feed their families. As a primate researcher and conservationist, I began to feel the weight of the world on my shoulders.

Over the next few months, I would learn that sojourning in Madagascar is like riding a roller coaster. Feelings of dismay and devastation can suddenly spin out into fascination, laughter, and joy. Now, as we made our way to Maevatanana, with the naked landscape an undeniable reality, I was about to experience this emotional roller coaster for the first time.

"Check it out," Shawn said, laughing. His voice startled me a little. I thought he was asleep. He was pointing out the window at a minibus that was rapidly approaching from behind our vehicle.

"It's a taxi brousse," he said animatedly and turned to me. "Is this the first one you have seen?"

I nodded. It was. And it didn't disappoint. I sat up in my seat and craned my neck to get a better view as the dark blue minibus whizzed past us. I could see that it was chock-full of people—so full, in fact, that it looked as though they couldn't close the door. A man was holding on to the sliding side door, his arm and shoulder exposed to the elements as the bus moved forward at an alarming speed. I also grew concerned when I saw just how much luggage was piled on top. I had been warned about the famous taxi brousse, but even so, I could not believe my eyes. The van was teetering unsteadily, boxes and bags piled so high that it looked as though the slightest gust of wind would cause the entire thing to come crashing down. Talk about top-heavy vehicles—Dr. Turner would not be impressed. I could not believe just how much luggage they had managed to load.

As we watched the bus disappear into the distance, Shawn turned to me again. "Hear that?"

How could I not? The taxi brousse stereo was on full blast, music blaring so loud that we would probably be listening to it for the next several kilometers.

I nodded. "They like their music loud, eh?"

"Salegy music."

"Salegy?"

"Yes," he said, and smiled. "It's the Malagasy national pop music. Famous. Heavy on the accordion."

To me, it sounded like fast, electric polka. It was a far cry from pop music back in Canada. This was no Taylor Swift or Justin Bieber. I would later read that salegy music originated in the north of Madagascar, and likely coincided with the arrival of the accordion in Madagascar in the nineteenth century.[20] In Jenny Fuhr's *Experiencing Rhythm: Contemporary Malagasy Music and Identity*, she turns to interviews with Eusèbe Jaojoby, a Malagasy composer and singer of salegy, to better understand the genre. He explains that before the accordion arrived in Madagascar, the musicians would perform a cappella—clapping their hands, or with drums.[21] Now, the electrified version of salegy was popular in nightclubs and at parties in Madagascar. Oh, and of course, blasting from the speakers of your taxi brousse.

As you make your way to the field site, the driver will often stop for lunch or dinner along the way. You should pay for everyone's food. The food at the local hotelys (small, roadside restaurants) is a wonderful way to sample the local Malagasy cuisine.

"Now's your chance, Keriann," Shawn said as we pulled to the side of the road for lunch. "You wanted to sample local food?"

We had pulled into a *hotely* restaurant just off the highway. I have to admit that I was feeling a little wary after my food poisoning in Tana. I pulled Shawn aside, "Is the food safe to eat here?"

"Don't worry," he said. "The food's fine as long as it is cooked. Just don't drink the water."

Andry and Sahoby led the way into the worn-down eatery and our crew dispersed among a few wooden tables. I glanced around. There was a chalkboard at the back, which appeared to be the menu. A few familiar words jumped out at me:

Hen' akoho, vary, trondro. Chicken, rice, fish. Andry had taught me a few Malagasy words when we were at the market in Tana.

"What would you like, Keriann?" Sahoby was about to place our order with the waitress. I ordered chicken and rice.

When the food came out, I could hardly believe my eyes. Imagine an order of chicken and rice in Canada. Do you see a dinner plate with a cup of cooked rice on one side and a six-ounce chicken breast on the other? Maybe a green salad to round out the meal? Chicken and rice in Madagascar is a bit different.

First, the waitress brought out the rice—a dinner-sized plate piled high. And I do mean high. Each plate had to contain at least four to five cups of cooked rice. These folks were serious about their rice.

I wondered if they had forgotten the chicken, but that came out next, on small side dishes, about the size and shape of a tea saucer. On the saucer sat a small drumstick and a couple of other pieces of chicken, along with about a quarter cup of broth. Finally, to round out the meal, the waitress placed a large bowl of extra broth in the middle of our table.

I paused to see how Andry and Sahoby proceeded. The rice was clearly the main event. The two students began by passing around the large bowl and spooning the broth onto their rice. Next, using spoons and forks, they pulled off small pieces of chicken, scooped up a bit of rice, and popped one small morsel at a time into their mouths. I mimicked their method. The food was good but there was no way that I could finish that giant mound of rice.

I had also hoped for a little more flavor. This food was bland. As a child, I had learned to love spicy food. My mom often cooked Indian, Mexican, or Thai food and was not afraid to add a little kick. While I was growing up, my father would often order the hottest options when we would dine out. Our family would have friendly competitions to see who could tolerate the spiciest meal.

Only once was my father ever defeated. It happened in Banff at the Magpie and Stump, a favorite Mexican restaurant for my family after a day of skiing at Lake Louise. My dad ordered the paella and the waitress asked him how he wanted it spiced.

"Hot," he said confidently.

The waitress raised her eyebrows, "Are you sure? Our hot is really, really hot."

Yes, he insisted. He could handle it. He was tough, he said. He had eaten spicy food in India, he had eaten it in Sri Lanka. But as he took the first bite, his eyes widened, and he reached for his water. He never did finish that meal. The rest of us had a good laugh, and now we remind him of it every chance we get.

Belize didn't disappoint for a spicy-food enthusiast. The local dishes were often on the spicy side, and to top it off the stores sold one of the best hot sauces that I have ever tried: Marie Sharp's. It is made with habanero peppers and available in all levels of spice from mild to "beware," and we grad students would buy bottles and bottles and douse our food in it each night. We would then line our suitcases with the stuff and bring it back to Canada. It was always a sad day when we polished off the last bottle of Marie Sharp's.

Now, as we dug into the *hotely* food, I asked Sahoby, "Is there a Malagasy hot sauce?"

"Sakay?" he asked. "Yes, we make sakay. Andry and I bought some for Kasijy."

"Oh, fantastic," I said and took another forkful of white rice.

Andry leaned forward. "When we get back to Tana I can take you for ravi-toto (ravi-too-too)." He explained that ravitoto consisted of shredded cassava leaves served with fried zebu or pork and coconut. I would later learn that ravitoto is a traditional and beloved dish among the Malagasy people—comfort food for many. In fact, on a visit to the Jumbo Score back in Tana, I would discover that you can purchase ravitoto in the can.

Among the *vazaha* researchers I know, there is a division between those who love ravitoto and those who hate it. Those who love it can get behind the idea that it is comfort food, and are happy to consume some greens, which are often lacking from other local dishes. And there is no shortage of greens in ravitoto. Those who hate it often have trouble with the texture. The cassava leaves are pureed, almost like creamed spinach but without the cream.

When I returned to Madagascar for my full field season, my cook would serve us ravitoto often—sometimes twice a day for lunch and dinner. I didn't mind at first, but by the end of the field season, I would dread the days that brought forth ravitoto. Yet for Andry, who grew up in Madagascar, ravitoto and rice was an important part of his food culture, and he craved it. For me, the dish was foreign, and left me yearning for some mac and cheese or lasagna. What would Andry think of Canadian food? I wondered. If he visited Toronto, would he miss the Malagasy staples he was used to?

After we had finished eating, Sahoby approached the counter and ordered snacks for the road—mofo ("moo-foo"). Now here was a food I could get behind. Mofo, or "bread," is the Malagasy answer to donuts. Sahoby ordered us two kinds—mofo baol and mofogasy. Mofo baol are round balls of sweetened dough, a little larger than a donut hole, and like a donut hole, they are deep-fried to perfection. Mofogasy are made with rice flour, water, sugar, and yeast and are baked in a mold, akin to a muffin tin. Both are delicious. Sahoby explained that these delectable treats are often consumed for breakfast with coffee or tea, and that these two kinds of mofo were merely representative. Mofo ananana are savory dough balls that contain chopped greens. Mofo akondro contain bananas. And mofo katsaka are made from sweet corn. I have my work cut out for me, I thought.

I watched the transaction. Sahoby handed over a 2,000 ariary bill and the clerk made change. But, unlike Canada where the change would come out of a register holding neatly stacked, organized, and relatively clean bills, the clerk here reached into a coffee can on the counter. She dug around in the can, which held several crumpled, darkened bills—money, but almost unrecognizable as such—and made change for Sahoby. Never before have I seen money as dirty as what she handed to Sahoby. I would later learn that it is normal in Madagascar to receive money and crumple it into a ball and stick it in your pocket for later.

I witnessed that behavior many times, especially in the small village markets, and I even found myself doing it near the end of my longer field season. There were no wallets or money clips around here. Aside from the money becoming worn out and torn in some cases, it also emitted a smell. On one memorable occasion, months after we had returned from a trip to Madagascar, Travis walked into his office.

"What's that smell?" he called out.

After some digging around, searching our backpacks for forgotten apples or dirty Tupperware, he figured it out: it was some leftover money he had brought back from Madagascar. After that, we began storing our leftover Malagasy currency in Ziploc bags.

I polished off my mofo and turned to Shawn, "Is there a washroom I can use?"

Any female traveler will tell you that making it through a long road trip in a foreign country is always an adventure. Men have it easy. On our way to the *hotely*, we had stopped a couple of times at the side of the road. While the men in our crew didn't have to go more than a few meters away from the vehicle, I found myself searching for a bush that was tall enough to provide a modicum of privacy—a bush that was sometimes quite a distance away. It had been a long drive for me in that way, filled with internal dialogue: *How badly do I have to go? Can I wait until there is something that resembles a washroom? Oh, here is a good patch of bushes, should I ask to pull over now? Oops, too late.*

Now, at the *hotely*, I knew that I should take advantage of a real toilet. Shawn directed me around the corner. When I arrived at the squat toilet, I hesitated. I thought, *Now, this is interesting.* It was quite literally a hole in the ground with two concrete rectangular slabs, flat on the ground on either side. The only thing private about the toilet was that it was around back. No walls shielded it. It was unclear to me how it worked. I had never had to use one of these before. Which direction should I face? I decided to go with facing out toward the restaurant, so that I could see what was coming. I didn't want to be surprised while I was indisposed. I slowly and carefully "assumed the position," when out of the corner of my eye I noticed that I could see people. If I could see them, I knew, they could see me.

What to do? Flustered, but already committed to the task at hand, I decided to ignore the prying eyes. I knew I would never see these people again. I also knew that, though they looked interested, they very likely didn't have any malicious intentions. Probably they were interested in looking because they had never yet seen a white female bum. And as we would drive into more and more remote communities, I knew that this kind of fascination would only increase. I did my business as quickly as I could. It was time to get back on the road.

The visits with the local officials on the way to your field site are vitally important to the success of your project.

We rolled into the Gendarmerie in Maevatanana at around three p.m.—right on schedule. All five of us—Shawn, Andry, Sahoby, Prospère, and I—piled into the office where the deputé of Maevatanana and the director-general of the Gendarmerie were waiting for us. Once again, Andry took the lead, translating for Shawn as needed. As expected, we would be required to bring along two gendarmes on our hike into Kasijy. They would stay with us until June 15, when they would head back out with Shawn.

"He says we must be careful," Andry turned to Shawn. "Because there may be bandits on the way to Kasijy."

Andry went on to explain that these bandits, or *malaso*, wouldn't necessarily be targeting us, but rather were out to steal zebu, Malagasy cattle, from local people who might be making their way between communities. If they did come across us, however, the officials were concerned that they might try to steal from us or harm us. They were especially worried about me, Andry said, the young white woman. My female-ness, it seemed, was becoming a common theme. I wondered why my being a woman was such cause for alarm. Was it because this was a male-dominant culture? Or, maybe because I was the only woman among a group of men? Or maybe I should be concerned? As the thoughts and worries swirled through my head, I was reminded of a conversation I had with Shawn before we left, after we had learned that Hilary wouldn't be joining us.

"What do you think about my being the only woman on this trip?" I had asked, meekly. "Will that be an issue?"

"Nah," he said. "My female graduate students have always done really well in Madagascar," he said.

"Oh?"

"They've been able to navigate local politics really well—better than I could—and have had no trouble in the field. And Malagasy women are tough, that's for sure—some of the strongest people I have met in my life."

Now, in the Gendarmerie, Shawn and I exchanged glances. His eyes told me not to worry. He looked back at the director-general and nodded politely, acknowledging the warnings. Then, in came the two gendarmes who would be joining us: Jean Paul and, funnily enough, another fellow named Andry. A common name, it turned out. The gray-haired Jean Paul was the older of the two, probably in his forties, while Gendarme Andry, a good-looking guy, was probably in his twenties, not much older than me. They were both dressed in green and beige camouflage military fatigues and carried AK-47 assault rifles. When they entered, fully kitted out, our trip suddenly felt a lot more serious.

Once the director-general had given our two newest team members their orders, we climbed back into the MICET vehicles and drove to a rickety ferry that would take us across the river to Mahazoma, where the deputé lived with his family, and where we would pitch our tents for the night. The ferry marked the last stop for the old blue MICET van and its driver, Pierre. In our meeting earlier, the deputé had confirmed what Shawn suspected—that the van would not be able to navigate the roads beyond this point. Before we crossed the river, we spent some time dividing up the gear that had been in the van between the other two vehicles. Then we waved goodbye to Pierre.

Gates surrounded the deputé's property. By Malagasy standards, he was extremely well-off. There was plenty of room for all of us to set up our tents (the MICET drivers would sleep in the vehicles, Sahoby told me). Inside, the house was immaculate, and the finished wooden floors sparkled. At dinnertime, we all crowded around a large, rectangular table for a meal of rice, chicken, and vegetables, which was decidedly less bland than the food we had eaten at the *hotely*.

It was fun to be invited into someone's home for dinner and a neat way to see how local people lived—albeit wealthy ones. Forging relationships with the local people in Belize had been a highlight during my master's fieldwork there. Every other day, we would hop into our small kayak and make our way to Monkey River Town where we would grab a drink at Ivan's and chat with locals. Travis would often play a round or two of pool with some of the guys while I made my way to Elna's house. We had been introduced to Elna by one of the other graduate students, Alison. Alison had put in some serious time in Belize for her master's and PhD research, and had forged strong relationships with many of the locals, especially Elna.

Elna welcomed all the graduate students into her home. We could drop by unannounced any time and she would be happy to sit and chat with us, filling us in on all the local gossip (it was a small town, after all). She would swap DVDs with us, and even taught me how to bake Belizean bread. Having Elna in Belize made my experience all the more special. As we sat around the deputé's table, I hoped that I could find similar relationships in Madagascar. Andry looked over at me and grinned, as though he knew just what I was thinking.

Satellite text message received from Travis Steffens
I miss you! You must be on your way to Kasijy now. Good luck in the field. I know you will do great.

CHAPTER 7

Testing the Young

May 31, 2006

The drive out to your field site is likely to be an eventful one. A good road in Madagascar is very different from a good road in Canada. But, not to worry, the MICET drivers are very experienced at navigating treacherous roads so you will be in good hands.

"Alefa, alefa," Sahoby called as he and the two gendarmes pushed the four-by-four from the rear. The vehicle's wheels spun, kicking back sand and mud. This was the third time one of the vehicles had got stuck in the mud. We were on our way to a town called Kandreho ("can-dree-hoo").

"I'm told the roads are decent," Shawn had said earlier that morning as we dismantled our tents. "When we get to Kandreho, we will have to meet with the local officials. That's where we can hire our cook and some porters to carry our supplies to Kasijy. I'm hopeful that today will be an easy drive."

Ha! Famous last words, I thought now as I watched the guys work to extricate the vehicle from the mud. We had become a team of nine: two drivers, two gendarmes, Prospère, Andry, Sahoby, Shawn, and me. Somehow, all nine of us had managed to cram into the remaining two four-by-four vehicles, which were already heavy with our supplies and sagged under our added weight. I sat in the back of a vehicle with Shawn up front, and next to me were Andry and Prospère. Sahoby and the two gendarmes went in the other vehicle. We had managed to get an early start and were on the road by eight a.m.

We quickly discovered that the "decent" roads were sandy, wet, and filled with unexpected patches of mud. And because we were now carrying so much weight, we kept getting stuck. Fortunately, we had experienced drivers. Each time we got stuck, they would jump into action, quickly calling out orders.

Everybody out. Andry, Sahoby—grab the slats of wood from the back of the vehicle (kept for just this reason). Wedge the wood under the vehicle's rear tires and get ready to push. As I watched the well-orchestrated spectacle each time we got stuck, I was transported back to my youth in Calgary, where the first big snowfall always came as a surprise, and getting stuck in the unplowed, snowy streets on the first morning was inevitable. I was interested to see that the techniques for getting unstuck from the mud in Madagascar were not that different from those we used in Calgary to escape the powdery snow.

But the mud wasn't even the worst of it. In fact, it was the least of our concerns. In between the sandy stretches of road, we had to navigate rocks—large boulders of sandstone. "This reminds me of the Hoodoos," I said to Shawn at one point as we stood waiting while the drivers worked out a game plan. The Hoodoos are oddly shaped rock formations, made of soft sandstone, and a distinct feature found in the badlands of Alberta, Canada, about 140 kilometers east of my Calgary hometown.[1]

The drivers faced a dilemma each time we came across a rocky area. Go over or go around? Can the vehicle clear that gap? They would stop, get out of their vehicles, and pace round and round, discussing the various obstacles, until they felt they had a solution. Then, one of them would get back in a vehicle while the other guided from outside. Then they would switch.

Every once in a while, the drivers would ask us to pile out of the vehicle while they navigated especially difficult sections. "We must get out," Andry would say.

We would all obediently pile out of the vehicle and wait for our signal to pile back in. More often than not I was grateful to get out and stretch my legs. It was certainly better than the alternative, especially considering my motion sickness. The times that we did stay in the vehicle, I sat in the rear seat and held on to the passenger door handle with a white-knuckled death grip. I couldn't believe that I had thought the highways were bad. What I wouldn't give to be back on the paved road, lurching slowly along the switchbacks. Now, I was getting a mean abdominal workout as we heaved through the treacherous landscape. The drive was exhausting—and I was just sitting there! The real kicker, though, was that not only were we navigating the challenging landscape, but we also repeatedly got lost on the so-called road. Between the various obstacles and, well, the lack of a road, we were forced to reverse out from dead ends that terminated in scrubby bushes on many occasions. And then, oops, stuck again.

"We must—"

"I know, I know," I would wave Andry off by the afternoon. "Get out."

It took six hours stopping and starting, extricating the vehicles from the mud, and piling in and out of our seats, to reach the town. We arrived around two p.m. We would stay at a spare house owned by the deputé, which he had kindly offered to let us use. After we had set up our tents on the property, Shawn and I went into the foyer of the ceramic-tiled concrete house to talk logistics. We sat down on the couch and, on the coffee table before us, Shawn spread out the laminated satellite images he had prepared.

"We're a little behind where I'd like to be with our timelines," he said. "We are going to need to stay two nights here instead of one so that we have time to meet with the local officials and organize our porters. But, as you can see here," he tapped the map, "we are now only about thirty-five kilometers from Kasijy."

"Will we be driving anymore?" I asked, holding my breath, and rubbing my abdominal muscles. *Please say no, please say no.*

"No, the MICET drivers tell me they can't go any farther." Shawn had hired a guide to help them navigate the drive back to the ferry. Andry and Sahoby were off-loading our equipment and supplies. "Tomorrow," Shawn continued, "we will ask the local officials to help us figure out the best way to get to Kasijy. Andry and Sahoby will organize our porters, who will help us carry our equipment and also find us a cook."

> *It is a good idea to hire a local cook while you are in the field. The cook will be responsible for organizing and keeping track of the food supplies and ensuring that all meals are available on schedule. Having a cook is an advantage in the field because your time and the time of your research assistants is better spent collecting data.*

In Madagascar, although it may seem elitist, researchers usually hire a local cook. This brings several advantages. First, it prevents the team from getting fatigued. We needed to spend our time and energy in the forest, collecting data. Having lunch and dinner ready when we got back to camp would make our time in the field much easier and would allow us to focus on the data. Second, hiring a local cook meant that we could more easily provide the team with the kind of food they were used to, and so keep morale high. The way that I prepare rice and beans wouldn't be the same as they do it in Madagascar, and I certainly did not know how to cook ranopango, the rice tea the locals were accustomed to drinking at the end of a meal.

I knew from experience that decent, calorie-rich food is important for surviving fieldwork. Aside from providing us with the nutrients we needed, having good food would contribute to our mental health and comfort while we were out in the bush. Finally, hiring another local team member as a cook also fosters good will with local communities and contributes to their economy. We would pay the cook a daily wage and provide him or her with a steady job.

Hiring a cook isn't without drawbacks, however. First, it means another person to manage and adds cost (albeit not that much) to the research budget. Given that the cook would very likely speak only Malagasy, Andry and Sahoby would do the managing. Second, we had to make sure that the cook knew how to cook for us *vazaha*. My experience with food poisoning back in Tana had underlined for me the importance of ensuring safe food practices. Andry and Sahoby would sit down with the cook and walk him or her through the basics, including hand washing and how to wash and dry the dishes safely.

Finally, hiring a local Malagasy cook meant that we *vazaha* might not get to eat our comfort foods. A researcher who had worked in Madagascar before me once told me a funny anecdote. She had been working in Ranomafana, in the eastern part of the country, and had hired a local cook. After about a month of rice, rice, rice, all she wanted was some pasta. So, she had made her way to a nearby store and bought a small bag of pasta to give to her cook. That night, her mouth watered in anticipation as the cook prepared her food. The cook proudly placed the dish down in front of her. There it sat: a giant mound of white rice, with a heaping portion of plain spaghetti noodles on top.

It will probably be necessary to overnight on your way to the field site. You will also need to visit various local officials along the way (e.g., Water and Forests, Gendarme). The student assistants are usually very good at negotiating with local officials about hiring porters, a cook, and guides.

The next morning, after a surprising and delicious breakfast of chocolate croissants that Prospère had coordinated (the French influence permeates the entire country, even remote towns), we made our way to a white, concrete government building in the center of town to meet with the mayor of Kandreho and the officials responsible for Kasijy Special Reserve. We wanted to introduce ourselves and ask about the best way to reach Kasijy.

En route to the government building, I got my first good look at a zebu, complete with cart. Zebu are ubiquitous in Madagascar. A subspecies of domestic cattle, they are thought to have originated in northeast India and later spread to Egypt, Ethiopia, and other parts of East Africa. Archaeologists suggest that settlers would have brought zebu to Madagascar at a very early date. Zebu are characterized by a fatty hump on their shoulders, like those on camels, and come in various colors. They weigh between three hundred and four hundred kilograms and average about 130 centimeters in height (just over four feet).

In Madagascar, zebu number more than half of the country's human population and represent a culturally important symbol of wealth and status. In

The Eighth Continent, Peter Tyson describes the importance of zebu as a status symbol: "Today, the net worth of entire villages," he writes, "may be wrapped up in cattle."[2] In fact, he points out, zebu are so culturally important that there are about eighty words in the Malagasy language to describe their physical attributes.[3] Shawn put it this way: "In Canada," he said, "you'll see men ogling women—you know, checking them out. Well, in Madagascar, I've caught men doing the same thing—the ol' elevator eyes? But when I look around, there are no women in sight—just zebu!"

Along with rice, zebu represent an important part of Madagascar's economy, which has led to what Tyson describes as a "cultural obsession."[4] The Malagasy use zebu for practical reasons (meat, milk, beasts of burden), but also accord them an important spiritual role in sacrificial ceremonies or funeral rituals.[5] Among the Sakalava, the zebu-ancestor bond is particularly strong, and zebu are used as ancestral sacrifices.[6]

In Tana, I had spotted a few zebu near the outskirts of town, and we had seen some as we navigated the main highway to Maevatanana. We had also eaten zebu meat while we were at La Maison du Pyla. It tasted like beef but was a little tougher, and certainly didn't hold a candle to Alberta beef, which I had grown up with. But now, in Kandreho, there sat two zebu immediately in front of me, attached by the nose to a two-wheeled wooden cart. The cart reminded me of something out of *Fiddler on the Roof*. It was filled with the familiar rice sacks that we were coming to know and love.

Much like rice, the zebu of Madagascar represent a complicated issue from a conservation perspective. To graze their cattle, the Malagasy often burn large patches of land to promote the growth of young grasses, and these fires can sometimes rage out of control and encroach on nearby forest patches.[7] Additionally, because the zebu are so culturally important, people will engage in more farming practices (such as maize cultivation), and thus more habitat clearing, as a way to earn money to purchase more cattle.[8]

The cultural significance of zebu and their role as a status symbol in Madagascar has created another issue as well: cattle theft. When we picked up our gendarmes in Maevatanana, the deputé had given us a serious warning about the very real possibility of running into gangs of thieves looking to steal zebu. These gangs, he explained, will sometimes wander the savannah near towns and communities, using armed force to steal cattle. These thieves are not shy to attack travelers as well, he said, which is why we were required to take the gendarmes with us. Who knew that innocuous-seeming cattle could have such an effect on our expedition, never mind the whole country?

Outside the government building, we encountered an older gentleman who reeked of booze. After some back-and-forth with Prospère, another man invited us inside where the reserve official was waiting to talk. There weren't enough seats, so Shawn suggested I could wait outside while he, Prospère, and Andry worked out the logistics. This left me standing with Sahoby and our drunken friend.

At first the situation was comical. The drunken man prattled on in Malagasy to Sahoby, laughing and gesticulating wildly. I couldn't understand Malagasy—especially not when it was slurred—so I watched Sahoby's face. At first it showed mild amusement—I mean, how else does one respond to a drunken stranger? But as the man continued, turning and gesturing to me a few times, I could see a shift in Sahoby's expression. The look of concern that he now wore, coupled with the way the man was eyeing me, made me very uncomfortable.

"What's he saying?" I asked Sahoby.

Sahoby was quiet for a moment and looked down at his feet. "Maybe you should go wait inside," he said.

I slipped inside and watched through the doorway as Sahoby tried to manage the situation. As I did so, my frustration mounted. Would a man have had to make such a retreat? Highly unlikely. But this was par for the course for a woman working in the field. In Belize, the local men had kept their distance from me because I had Travis there as a buffer. But even that buffer wasn't 100 percent foolproof. I remember one day during my master's research. We were paddling home from Monkey River Town when we came upon two young boys, about twelve or thirteen years old, in a boat. One of them called out to us: "Hey, Travis! I like your girl!"

We laughed it off—they were just kids, after all. But that kind of catcalling was a symptom of the unfortunate reality that women, travelers in particular, were often objectified. Objectification was easily dealt with. The greater concern was that simple objectification could easily and quickly become something more serious. Situations like that, and like the drunken man now in Kandreho, added stress that male researchers didn't have to worry about. Thankfully, Sahoby navigated this situation like a pro, and did not allow it to escalate any further. I was grateful that he had been there, and it struck me: *What if he hadn't been? What would I have done?*

After the meetings, as Shawn and I walked back to the deputé's house, I debated whether to say something. On the surface, nothing had happened. But something about that situation didn't feel right—it wasn't sitting well. *Should I tell Shawn?*

Everyone who is close to me would tell you that I am not an over-sharer. In fact, some might say that I am the opposite. I have been "the quiet one" my entire life. A classic introvert. When I was a teenager, my parents used to tease me: "It's so strange, Keriann," my father would say, a twinkle in his eye, "that there aren't any boys in your class at school." My mom would burst out laughing and add: "That's right, Ken. We only ever hear about the girls." The two jokesters would chuckle while my face grew red. But my tendency to keep things to myself started long before I was a teenager.

The first time I remember actively concealing information—important information—was when I was five. I was in kindergarten and my brother, Carlin, was in grade two. We both went to the same elementary school. I looked up to Carlin. In my eyes, he was a legend. I lived for recess because it meant I might get a Carlin-sighting. Seeing my older brother across the school yard was akin to spotting a movie star. One particular day, I stood with my friend Lori near the tetherball pole (remember tetherball?), and my brother was a few feet away with his friends. He knew I was nearby but paid no attention. He couldn't be seen associating with his younger sister, after all. As Lori and I stood snacking on our miniature boxes of raisins, an older girl approached. She was maybe ten or eleven. I smiled at her. A new friend perhaps?

BOOOSH!

Out of nowhere, the girl punched me right in the stomach. It was the first time that I can remember feeling winded. Even as an adult, getting the wind knocked out of you is staggering. Enough to bring you to your knees. As a five-year-old, the first time this had ever happened in my life, I was convinced that I might die. I bent over and grasped my stomach. I couldn't speak. What had just happened? I looked to my friend, Lori. She was wild-eyed, no doubt struggling to process the sequence of events.

Then it dawned on me. I knew what I had to do. I rushed over, still winded, to where my brother stood. In tears, I got the words out: "Th-that girl," I pointed, "j-just punched me."

Carlin didn't hesitate.

"Hey!" he yelled and rushed toward the girl. Upon seeing my brother, the mysterious older girl (I still don't know where she came from!) made a fast retreat, hopping the school-yard fence. My brother the hero! Talk about solidifying his superstar status in the eyes of his younger sister. The bell rang, and we all filed into our respective classrooms.

I remember sitting on the rug that afternoon as my teacher, Mrs. Field, taught us about colors. I knew about colors. In fact, I could feel the color rushing to my face. I had a secret. Someone had just punched me—hurt me. And yet . . . I said nothing. My five-year-old brain half expected an adult to ask me about what had happened. But, of course, no one did. How would they know to ask? Only Lori had watched the event unfold. The school day ended and Carlin and I walked home. When we got back to the house, my mom had readied an afternoon snack for us. As we sat down at the table to dig in, my mom asked how our day was. I kept quiet. Did she know what happened, I wondered. She didn't seem to. Carlin looked my way, perhaps waiting for me to tell the story. When I still said nothing, he gave me a strange look and told my mom every detail.

To this day, I am not sure exactly why I decided to keep quiet. I was only five, so part of it, surely, was that I didn't have the words to express what had happened, or the emotional bandwidth to process it. The other part, though, I think, is a desire not to trouble other people. To keep things—even bad things—to myself, to deal with them myself, unless I absolutely have to tell someone.

Now, in Kandreho, as the government building grew distant, I could feel a giant knot growing in my stomach. *Should I say something?* I couldn't shake the growing feeling of violation. I decided to fill Shawn in. As I told the story, I found myself getting worked up and emotional. The feelings that now rushed to the surface surprised even me. The words tumbled out. My face flushed. My eyes watered. But physical reaction aside, what was even more surprising was that I also found myself questioning these emotions. *Maybe I am overreacting,* I thought. *Why am I so upset?* I felt embarrassed to tell Shawn about what had happened. Worried about his reaction. Would he brush it off as a non-event? Would my telling him make him feel uncomfortable? But, wait, I would catch myself. Why am I worried about Shawn's feelings? What does this whole incident, and my reaction, say about me?

To his credit, Shawn didn't brush it off. His eyes widened as I described the scene.

"Keriann, I had no idea." His voice was quiet, and he was visibly disturbed. "If anything like that happens again, come and find me right away."

"Sahoby had it under control," I said. I caught myself for a moment. For some reason, I was unable to stop myself from downplaying my feelings. *Why was that?*

"Thank goodness. At any rate, you do not have to ever go back there," Shawn said, a serious look in his eyes. "I mean that. You should not be made to feel uncomfortable."

Shawn's sincere concern and support was exactly what I needed. In that moment, like my brother when I was five, Shawn was my hero. He recognized that my feelings were valid. I had every right to be upset. *I should not be made to feel uncomfortable.* Those eight words, somehow, made me feel a million times better.

After a brief pause while we both digested what had just happened, Shawn broke the awkward silence by filling me in on his meeting with the reserve official. He relayed both good news and bad. "As we thought," he said, "it's about a thirty-five-kilometer hike from here to Kasijy, and it sounds doable, which is good. I asked Andry and Sahoby to go and find some porters and a cook. Once they come back, we will make the final arrangements. We should try to leave early tomorrow so that we don't have to hike in the heat. But the bad news is that they are requiring that we bring a *third* gendarme with us from Kandreho. I did what I could, but I couldn't convince them otherwise. They are really concerned about bandits and said they need three men for strategy . . . whatever that means. So, when they go and get the porters, Andry and Sahoby are going to stop at the market and pick up some more rice and beans to accommodate another person." Shawn suddenly stopped walking. Something had caught his attention. He pointed, "What's going on there, I wonder."

Down the road, a large crowd of people had gathered. Music was blaring—salegy, I remembered—and we could hear shouting, cheering, and laughing. We drew closer to investigate. Here, Shawn's height came in handy because he could easily see over the crowd.

"It looks like some kind of . . . wrestling match?" Shawn looked back at me and smiled. "Let's check it out."

Indeed, it was some kind of game. We stood for a while, watching and trying to work out the rules. We distinguished two teams of boys and young men, ranging from ten to twenty or even thirty years old. There also appeared to be a few guys in charge, tasked with deciding who would fight whom. Once the boys were paired off, the wrestling began. That's where the semblance of rules broke down. The boys moved into the center of the crowd where they proceeded to fight. It seemed as if almost anything was acceptable—they punched, kicked, scratched. We watched three or four pairs of boys go into the "ring," but we still couldn't figure out how the winner was determined. Each time, the fight would just stop abruptly.

This scene intrigued me so much that later, when I got back to Canada, I went down a Google rabbit hole until I finally found an excerpt describing *moraingy*, a Malagasy fighting style. That was it. Moraingy is a weaponless, bare-fisted fighting style that originated during the Maroseranana dynasty of the Sakalava Kingdom. It evolved as an important element of male culture among the Sakalava in which the elders can test the strength and abilities of the youth.[9]

Young men are keen to participate because they get a chance to demonstrate their strength and gain prestige in their village. Moraingy matches are described as having an atmosphere of "mutual respect" and "loyalty," solidifying cultural identity in young men, providing physical education, and promoting cohesion between neighboring villages. In a typical moraingy match, as we witnessed that day in Kandreho, the elders will pair up two fighters—usually two from different villages. There is only a single round, when the two boys punch and kick, and the round ends when one of the fighters voluntarily or accidently exits the arena, faints, or is clearly unable to defend himself any further. The judge—one of the elder men—declares the victor.[10]

Shawn and I were enthralled.

"This is incredible," Shawn said. "In all my time in Madagascar, I've never seen anything like this."

After about twenty minutes of watching the moraingy matches, we finally pulled ourselves away and headed back to the deputé's house, where Andry and Sahoby were waiting. They had found us a cook, a young man named Fidèle. Hiring porters, however, had proven more challenging. Andry explained that they couldn't find enough men willing to carry our gear out, and so instead they had arranged to hire a combination of porters and two men with zebu carts. The carts could hold about 200 kilograms, and so could transport more gear than a single man could carry.

The drawback of using zebu carts, Andry said, is that they traveled so slowly. He estimated that what we had projected as a one-day hike would now take us two days. The first twenty kilometers or so would take the carts about eight or nine hours, because the zebu needed to stop periodically to be rested, fed, and watered. And another catch: the carts would only be able to come with us for that first day. After the initial twenty kilometers, the zebu cart road ended, and so we would need to camp for one night, off-load our equipment from the carts, and try and find some porters from a nearby village to help carry our gear the remaining fifteen kilometers to Kasijy.

"Wow," Shawn said after Andry had finished. "That is a lot more complicated than I thought."

It was troubling to hear that Shawn hadn't expected the hike to be so complicated. I was relying on his expertise. Should I have done more research myself? I tried to remain positive. "Well, at least we will be on our way tomorrow," I offered.

"There is one more thing," Andry said, and glanced over at his fellow student. "Sahoby is sick."

Satellite text message received from Travis Steffens
I got your message. Why do you need armed guards? Too bad about Sahoby. Hang in there.

CHAPTER 8

A Sore Throat, an AK-47, and a Useless iPod

June 1, 2006

As it turned out, Sahoby's beautiful silk scarf and his World War II pilot-look were for function, not fashion. Sahoby had been battling a severe sore throat since before we left Tana, and now that we had arrived in Kandreho he was feeling worse than ever.

"The scarf helps soothe my throat," he explained. This was a remedy I had never tried, but okay.

Shawn looked concerned. "Have you been to the doctor?"

"Not yet," said Sahoby, "but there is a doctor here in Kandreho. I can go this afternoon."

"Yes, please go as soon as possible. Why didn't you tell us this before we left Tana?" Shawn didn't wait for a response. "Next time don't wait to tell us if you are sick or hurt. The health of everyone on our team is very important. We're all in this together."

I took in Shawn's words. He was right. We were a team. And it was true: Sahoby should have told us earlier that he was unwell. But at the same time, I could understand Sahoby's motivations for hiding this piece of information. It had been no secret that we were behind schedule due to Shawn's and my food poisoning. Our departure had already been delayed by several days, and Sahoby probably figured—hoped at least—that he would simply get better after a few days. He probably didn't want to be the reason for our trip being even more delayed—Shawn's departure date loomed heavy over all our heads. And so, Sahoby had kept quiet. But, as our adventure unfolded, he was forced into physical challenges, such as pushing the four-by-fours out of the mud. It was no surprise that his condition had worsened. Now, he couldn't hide it anymore.

I knew from experience that it was crucial to attend to illnesses like Sahoby's while in remote areas and I hoped that the doctor in Kandreho would be able to provide effective medication. I had suffered through a dicey experience back in Belize. Travis was in Monkey River working as a field assistant to our friend Greg, while I was back in Calgary, analyzing data and writing up my master's thesis. But we had planned for me to visit Belize partway through his time there, in October.

When I arrived, the two guys were in full bachelor mode. They would regularly throw things at each other, laugh at each other's injuries, and don't even get me started on the dirty dishes. Greg was living up in the main researcher cabin, and Travis had bunked down a few hundred meters away, in a smaller house next to the river. The day after I arrived, while we all sat upstairs catching up, Greg had stubbed his toe badly on his way to the bathroom.

"Tsk, aww," I said, instinctively. "Are you okay, Greg?"

Greg looked up, wide-eyed. "That's it," he said. "That's what we've been missing for the past two months. The 'tsk-aww' factor!"

Travis quickly agreed. I just shook my head in disbelief. Clearly these two jokers had been without female sensibilities for far too long.

A few weeks later, the "tsk-aww" factor came in handier than I would have liked. A bad cold was making its way around Monkey River village and it hit Travis hard. He had to take a few days off to rest while I helped Greg in the field. Eventually, we decided that the cold was bad enough that we should get Travis to the doctor. We hitched a ride with a local friend and, on his recommendation, visited the public clinic in Independence.

The clinic was full of people who didn't look well. We patiently waited our turn, and when Travis saw the doctor, the whole visit started to feel like a mistake. The man took a cursory glance at Travis, peeked in his mouth for a few seconds, and then shuffled us out the door with a hastily written prescription to the pharmacist. The woman at the pharmacy window squinted at the prescription and divvied up a few different-colored unmarked pills, and handed them to us in a plastic sandwich bag.

Travis held up the bag. "What do you think these are?"

"No idea," I said. "But maybe they will cure you?"

They didn't. In fact, back in the cabin, Travis's condition worsened. He was unable to go into the forest and was fighting a bad fever. Greg and I were getting increasingly concerned.

"If he's not better by tomorrow, I want to take him to the doctor again," I told Greg as we walked back from our day's work. "But this time the private doctor, Pedro, whom I saw when I had my wisdom teeth problem."

When we got back to camp, Travis put on a brave face and a smile to greet us, but we could see that he wasn't himself. He had a bad fever, and his throat still hurt. We agreed that the next day we would go see the doctor. That night, I was deep in sleep when Travis suddenly shook me awake.

"Keriann!" he whispered urgently. "Keriann!"

"What is it?" I startled awake. Was he okay? My mind reeled—did I need to figure out a way to get him to a hospital?

"Wake up!" he said. "There's something outside." Travis was lying on his stomach, his chin resting on the windowsill at the head of the bed. He was peering out the window.

Oh dear, I thought. He must be so feverish that he's hallucinating. I reached over to feel his head and he pushed my hand away.

"No, look—do you see it?"

"I don't have my glasses on, Travis. You're sick, you should try to—"

Travis reached over and, with shocking speed and agility, snatched up my glasses case and handed it to me.

I sighed. I guessed I would have to play along. I put on my glasses and looked out the window.

"You see? Right there," Travis pointed to a darkened, unmoving object outside. To me, it looked like a snapped tree trunk. To him? "It's a—it's a monster."

Whoa. This was crazy. Travis? Saying he's seeing a monster? Travis is the most pragmatic, grounded, logical person I know, and would never in a million years believe that he could be seeing a monster. He was sicker than I had thought.

"Travis, it's not a monster," I said, trying to stay calm. "It's a dead tree. Don't worry. Now, you have got to try and get some rest. Let me get you some Tylenol and a cool cloth."

I managed to get some Tylenol into him, but for a good twenty minutes he sat there, his head perched at the windowsill, worried that this monster was coming for us. After about an hour, his fever broke. I could feel him start to sweat. In the morning, after the sun came up, I looked out the window as Travis slept on. It wasn't a dead tree after all. It was a stack of old tires that I had forgotten about.

On the drive out to Independence, I relayed the story to Greg, and we couldn't help laughing at poor Travis, who had now come to his senses. It turned out that Travis thought the monster from the movie *The Ring*—Samara—was crawling out of the well (aka the stack of tires). He was embarrassed and knew he would never live this one down.

When we got to Pedro's, we handed over the baggie full of pills that the doctor at the public clinic had given Travis. Pedro poured the pills out into his hand and held them up one by one. "This? This is acetaminophen. For headaches and fever." He threw that pill into the garbage and held up the next one. "This one is an antihistamine. For allergies." He threw that pill into the garbage. "And this," he said, smiling. "Well, this one, I don't even know what it is. Let's get you some antibiotics."

In the end, Travis got two shots of penicillin in his butt (with brand-new, sterilized needles straight out of the package). Within five minutes, he felt like a million bucks. If only we had gone to Pedro in the first place.

Now in Kandreho, Sahoby returned from the village doctor, medication in hand. We had a team meeting. Shawn decided that we would wait one more day in Kandreho so that Sahoby could start his medication and rest before the long hike. A wise decision, I thought—my experience with Travis had taught me that not all medications are created equally. The rest of the plan remained the same; we were just delayed another day.

"Andry, can you arrange for the zebu carts and drivers to spend tomorrow night near us here so that we can load them early in the morning? Ideally, we should start walking by seven a.m. at the latest so that we can beat the heat."

Did I mention that northwestern Madagascar is a lot hotter and drier than Tana? That city, situated in the central highlands, had been quite comfortable, in fact. It was sunny and warm during the day and cooled right down at night. But the island of Madagascar, which is only slightly smaller (at 587,041 square kilometers)[1] than the province of Alberta or the state of Texas, has an incredibly diverse landscape and an equally variable climate. Madagascar is in the subtropics, but because of the island's north–south orientation and eclectic geographic profile, there is immense climatic variation.[2] The mountain range that runs down its middle divides the country in two. East of the mountain range lie the rain forests. In Kasijy, on the other hand, we would find dry forests. We were also already learning that temperatures here were hot—no fooling. In the heat of the day in Kandreho, we had seen the thermometer rise above thirty-five degrees Celsius. It was all we could do to stay cool. Thankfully, the deputé's house was outfitted with fans and the courtyard shaded with tall, leafy trees. In the afternoons, Shawn and I would pull out chairs into the courtyard and read our books in the shade. So far, the heat hadn't been an issue, but now we were facing a hike. No more retreating into the shade. I thought of my wide-brimmed hat.

The day we now spent in Kandreho passed by slowly. We were all getting antsy, of course. We wanted to hit the road. Andry, Sahoby, and the gendarmes passed the time sitting in the deputé's house, shooting the breeze. When I would join the group, Andry would kindly translate for me periodically, but I still missed out on large portions of the conversation and couldn't help feeling a little left out. Luckily, all the guys took pleasure in listening to the *vazaha* girl practice her Malagasy.

Malagasy language, like the culture, is Malayo-Polynesian in origin, and most closely related to the languages of the Ma'anyan of Kalimantan, Indonesia.[3] In his book *The Eighth Continent*, Peter Tyson explains that the written Malagasy language, developed by missionaries and later adopted by King Radama

I, is a hybrid, with spelling combining Malagasy, English, French, and Arabic elements.[4] He writes that when "Welsh missionaries adapted the Latin alphabet to the Malagasy language in 1823," they tried to appease the competing English and French by stipulating "that the consonants would have their English values and the vowels, with two exceptions, their French ones."[5]

The two exceptions, I had already learned, were the terminal *y* and the letter *o*. Tyson explains that "when missionaries put Malagasy to the Roman alphabet they borrowed the Romanized spellings *ao* for the 'o' sound (as in 'low') and *o* for the 'oo' sound (as in 'loot') from Arabic."[6] Thus, the *o* in Kandreho was pronounced "oo." The Malagasy alphabet contains twenty-one letters, omitting the letters *c*, *q*, *u*, *w*, and *x*. Words always end in a vowel, which is usually silent.

As Sahoby gave me a quick grammar lesson, I learned that Malagasy is a forgiving language. Plurality simply doesn't exist, for example, and is inferred contextually as opposed to through the morphology of the words themselves. And unlike French, there is no need to worry about grammatical gender. For example, Tyson explains, *izy* can mean he, she, it, or they.[7] Sahoby explained that to conjugate a verb from past or present to future tense in Malagasy requires only a change to the first letter of that verb. For example, the verb "to eat" in its present form is *mihinana*. In the future tense, you replace the *m* with an *h*: *hihinana*, "will eat." To move into the past tense, just swap that *h* with an *n*: *nihinana*, "ate." Finally, negation is expressed by placing *tsy* in front of the verb.

Understanding simple negation in Malagasy is handy when you have a limited vocabulary. That's a lesson I would learn two years later, when I returned to Madagascar with Travis. It was the beginning of the rainy season, December, which meant wasps. We had learned that there are several types of wasps in northwest Madagascar and that they permeate the forest in the rainy months. We could differentiate the wasps by color—classic yellow, green, and black—and each, I would learn, had a sting that felt different from the others. I am not the most graceful creature on earth, and when I move through the forest, I have a bad habit of moving branches and leaves away from my face using my hands. Couple that with my need to keep a constant eye on the lemurs, rather than the forest in front of me, and you have a recipe for disaster. The wasp's nests are small and well hidden in the forest, often hanging beneath branches or leaves. Almost every day in December and January, wasps would sting me at least once, if not numerous times.

And so, one day when Travis was walking the trails, making a trail map with his GPS along with one of our guides, the two came across a small wasp's nest hanging alongside one of our trails. Travis was aware of my constant battle with wasps and so, gentleman that he is, decided that he should remove the threat completely. He decided he would cut the nest down. His Malagasy was still rudimentary, but he did his best to tell the guide what he was about to do. Travis pulled out his trusty pocketknife and began slowly approaching the nest.

"Mora, mora," he instructed. Slow.

Whack! With a flick of his wrist, he chopped down the branch holding the nest. Travis knew they needed to run. But suddenly, he realized he had forgotten the Malagasy word for run, and he didn't know how to say fast.

"Tsy mora!" he exclaimed, abruptly. Not slow! The two of them hightailed it out of there at speed.

Negation at its finest.

Another nifty trick to learn about the Malagasy language is that one word is often appropriated in several contexts, so learning one word can often open up an understanding into many words. For example, the word for salt in Malagasy is *tsira*. The word for sweet is *mamy*. So, what's the word for sugar, you might ask? *Tsiramamy*.

Listening to someone speak Malagasy is calming because the flow of words is almost musical. But Malagasy words can be tough for *vazaha* to wrap their heads around. The words are often lengthy, containing several syllables and many vowels. Tyson plays with word length in one of my favorite passages from his book:

> As you can see, Malagasy words can wind up being a mouthful, even if they're just a bite (*ambilombazana*) or even a crumb (*sombint-sombiny*). A stroll (*fitsangantsanganana*) to the town hall (*tranom-pokonolona*) at midday (*antoandrobenananahary*) with a member of parliament (*solombavambahoaka*) might entice (*fanambatambazana*) or even dazzle (*mampipendrampendrana*) you, though it might leave you at a loss for words (*miambakavaka*).[8]

After my language lesson was done, Andry and Sahoby began peppering the two gendarmes with questions. The students were fascinated with the military. "What do you do every day?" they asked, wide-eyed. I knew that the national gendarmerie is a branch of the Madagascar armed forces that is tasked with maintaining law and order in rural areas. There are about 8,100 gendarmes in Madagascar, and according to a 2011 Small Arms Survey published by Cambridge University Press, the gendarmerie is a tangled political web "characterized by an inflated proportion of high-ranking officers, a meddling in domestic policies, and entrepreneurial enrichment."[9] Shawn was aware of the political mess that was the gendarmerie in Madagascar. He had worked in the country for years. His hesitancy with bringing the gendarmes with us to Kasijy, I knew, probably had more to do with the political nightmare that could ensue, as distinct from the simple matter of the extra expense.

The word *gendarmerie* is derived from a medieval French term, *gens d'armes* ("armed men"). True to their title, our guards were armed. They were outfitted in military fatigues and carried AK-47 assault rifles. I watched now as Andry and Sahoby fawned over the gendarmes. The two Tana students were especially captivated by the rifles. Growing up with an older brother, I had been exposed to the odd fascination young men have with weapons. It was something I couldn't even pretend to understand. I supposed it had to do with the promise of the power that came with a gun, but that was only a guess. As we sat in the deputé's living room, all of us bored with the long wait, the two students asked gendarme Andry to show them his gun. The guard obliged and laid his gun out on the coffee table. He talked animatedly in Malagasy, as he pointed to the various features of the gun. Andry—the student—asked a question that made the guard hesitate.

"What did he say?" I whispered to Sahoby.

"He wants to hold the gun," Sahoby replied.

"Ohhh, I don't think—"

But before I could get the words out, the guard picked up the gun and cocked it, making the unmistakable sound we have all heard in movies and on television. He made to hand the gun over to Andry, when I heard a crash from outside and Shawn came rushing in.

"What's going on?" Shawn said, his eyes darting from me to Sahoby, and finally to Andry, whose arms were outstretched. I had never seen Shawn look so frantic. He must have heard the sound of the gun cocking and thought that we were in grave danger.

Sahoby was the first to speak: "He was just showing us the gun. It's not loaded," he offered.

Relief washed over Shawn's face, replaced quickly by a look of annoyance. He turned to the guards. "Please don't play with your weapons around my students," he said firmly and then walked back outside.

I stood and dashed outside to the courtyard where Shawn was sitting in the shade, clearly collecting himself after the scare.

"Sorry about that," I said quietly. "It all happened so quickly, and they were speaking in Malagasy. But still, I should have said something to them earlier."

Shawn, I could see, was becoming antsy and irritable, and the rifle incident had only pushed him further over the edge.

"Yeah, probably," he said. And then his voice softened: "It's not your fault."

I stood next to him and changed the subject to our journey ahead. We spoke of the various delays we had experienced.

"It's so frustrating," he said. "Because we are so close. Just thirty-five kilometers away. The more we are delayed, the less time I will have in Kasijy. I hope we can leave tomorrow. If we don't, I'll have only a few days in the forest before I have to turn right back around."

"I'm sure it'll be fine," I said.

Shawn rolled his eyes. "Well, if you're *sure*," he said, his voice sarcastic.

This awkward exchange reminded me of something that was said at my orientation for the University of Toronto PhD program. During a panel discussion titled "Making the Most of Your Graduate Experience," one of the panelists, a School of Graduate Studies administrator, discussed the supervisor-supervisee relationship.

"Here's an interesting way to think about it," she said. "The time it takes PhD students to complete their programs—typically about five to seven years—is longer than many marriages last. You are about to enter a long-term professional relationship. You'll hit good times and bad, just like a marriage."

Shawn's moodiness conjured that analogy. We were having a rough patch. But if this was as bad as it got, I reasoned, things were pretty good. As I recovered from the sting of Shawn's words, I reminded myself that he was an excellent supervisor—always responsive and supportive of me on both a personal and a professional level. This was just a high-stress situation—no doubt about that. Shawn's impatient tone was likely because he was worried about leaving me in Kasijy—a young female *vazaha* alone with a handful of men in remotest Madagascar, where lawless bandits were known to roam. This was not an idea I was ready to face, because if Shawn was scared, what should I have been? I slowly sidled away and went back into the house where the other guys were sitting. I would just keep practicing my Malagasy.

If you require porters to help with your field equipment, I suggest using a numbering system where each porter selects a number from a hat, and then they collect the corresponding sack. This method will make it easier to pay the porters at the end, especially because it is common to pick up people along the way. Once the sacks have been divided amongst the porters, you can proceed to your site.

I read through the section on porters in Shawn's unofficial guide that night—June 2. The men with their zebu carts had arrived as requested, and we were going to pack up and head out early next morning. We went to sleep for one last time in our tents in Kandreho. This was it. Tomorrow we would begin the hike into Kasijy. I couldn't wait. I had been in Madagascar for eleven days and I had yet to see a lemur. But now the lemurs were within spitting distance—a mere thirty-five kilometers away.

That night I prepared and organized all my equipment for our departure. I made sure that I had a rice sack with room to slip in my tent, sleeping bag, mattress, and extra clothes. Next, I prepared my day pack. I took two bottles of water and filled up my camel pack bladder. I knew that I would need a lot of water for this hike. I slipped the full bladder into the pack, threaded the hose through the top, and attached it to the strap. At the bottom of my pack, I carefully placed my plastic bag full of ariary and my passport. I added my binoculars and camera. I snuck in a snack, just in case—a peanut butter Cliff bar. At the top, I put in my water filter and an empty Nalgene bottle—the water I had just used to fill up my bladder would be the last bottled water that I would have. I zipped up my now full pack, and then noticed my Teva sandals. I would need to bring those as well, in case we had to go through any water or mud. I slipped them into the front pouch of my pack and attached them with a carabiner. My pack now full, I turned to my next chore. I needed to have everything ready so that I could get up and go. I laid out my clothes, found my hat and sunglasses and added them to the pile, and then I tucked the socks I would wear into my boots. There. All set.

We awoke bright and early next day to find that the locals had packed up their carts in the middle of the night and decamped.

"What happened?" Shawn asked Andry.

Andry just shook his head. "I'm not sure."

After a quick breakfast, Andry and Sahoby set out to find the porters. They returned at around 10:30, with two zebu carts kicking up dust behind them. The original zebu cart drivers had quit. Andry and Sahoby had managed to find two new drivers with carts who were willing to take us.

"Wow," Shawn marveled. "This is such a different experience than in the east. There, we had the problem of too many people wanting the work."

In *The Weight of the Past: Living with History in Mahajanga, Madagascar*, Michael Lambek unpacks Sakalava culture by exploring the historical roots as well as contemporary Sakalava perceptions of themselves. The Sakalava, he says, self-identify as *miavong*, meaning "reserved, standoffish, or proud."[10] Lambek has even heard some Sakalava describe themselves as "lazy," because they prefer fishing or collecting shellfish to the labor of growing rice.

In Nosy Be, an island off the northwest coast of Madagascar, Peter Tyson recounts similar difficulties in getting buy-in from the locals. The villagers "couldn't comprehend why this ragtag group of Malagasy and vazaha would come so far to collect small animals,"[11] he writes, and the team could only find one resident to sign on as a porter. Their best guess was that the reluctance to work could have been related to the fact that two weeks earlier the villagers had found and killed two cattle rustlers, and so were worried for their safety.[12]

But really, it was anyone's guess as to why we were having difficulties with porters. Sure, it could have been related to the Sakalava culture around work. It could have been an underlying feeling of skepticism about us *vazaha*. It could have been a perception that where we were headed was dangerous. It may even have been that it was now winter in Madagascar and the early mornings, when we were looking to leave, were cold. Whatever it was, it was new to all of us and we were navigating it as best we could.

We left Kandreho at 11:30 a.m.—in the heat of the day. We were a real crew now—three gendarmes; Shawn; Andry; Sahoby; our cook, Fidèle; the zebu carts and drivers; a handful of porters . . . and a partridge in a pear tree? I strapped on my day pack and readied myself for the hike. I was antsy from all the waiting around and raring to go. The porters led the way with Shawn close behind. Though I was ready to hit the road, I wasn't quite prepared for just how quick the pace was. Nor was I prepared for how hot it was. As we walked, I could feel the sun blasting down on my head and shoulders, and I thought about how grateful I was to have my unfashionable wide-brimmed hat. I fell back and kept pace with Andry, the two of us suffering from the heat, it seemed, in equal amounts.

For the first fifteen minutes, as we wended our way from the deputé's house to the outskirts of Kandreho, our route followed the hard-packed dirt roads. But soon, we neared the shores of the Mahavavy River and began following its course, our feet digging in and out of the soft, dry sand as we walked. Shawn had warned me that the terrain would be tough. But I didn't think it would be this tough. My mind had gone to hard-packed trails through shady forest, not loose sandy riverbeds where we would be fully exposed to the elements. Anyone who has walked along a beach knows the challenges involved with walking on sand.

It is so challenging that scientists have published articles about it. One article from 1998, titled *Mechanics and Energetics of Human Locomotion on Sand* (yes, that's the actual title), explored the mechanical work humans perform during walking and running on sand versus a hard surface. The verdict? Walking on sand requires 1.6–2.5 times more mechanical work than walking on a hard surface at the same speed. Likewise, walking on sand requires 2.1–2.7 times more energy than walking on a hard surface at the same speed.[13] In other words, sand is way more difficult! I didn't need the science to tell me that as we slowly plowed forward toward Kasijy.

I had hoped that I would be able to chat with Shawn as we hiked, but he was way up front now, propelled by his long legs and plugged into his portable CD player. I watched him, full of envy, as he jammed to his tunes. I had brought an iPod, sure, but I hadn't thought about how I would charge it in the bush. I had only realized that it would be impossible to charge it in Kasijy—there was no electricity, of course—when I learned that Shawn had opted for a battery-operated music device. I felt foolish now. Shawn had known that there would be no electricity, and he had known that the hike into Kasijy would be a mental game. He had brought a battery-operated distraction. *Why didn't I think of that?*

Back in Toronto, I had been a part of a running club, training for a half marathon that I completed one month before I left for Madagascar. Each week, my running group met to learn more about becoming a better runner, and one day the group leader arranged for a sports psychologist to talk to us. The young, fit, female psychologist told us that long-distance runners use different psychological strategies to get them through the toughest moments. *Dissociation* is one of those strategies. Athletes that dissociate are looking for a distraction—basically thinking about anything but the task at hand. There are ways to dissociate including internal reflection, counting cars, or listening to music. When I trained for that half marathon, dissociation—particularly music or audio books—was my strategy of choice. But now, in Kasijy, I had failed to bring a battery-powered music device. I thought I had been wise to bring a video iPod—I could watch movies—or so I thought. With no way of charging it in the bush, my iPod had become a useless paperweight. I had charged it as much as possible, and now that charge would have to last me the entire three months. I had to ration. I vowed to pull it out only at my most desperate moments.

With no music available, I thought now might be a good time to try out another psychological strategy the speaker had told us about—an *associative* strategy. *Association* is a strategy often implemented by elite marathoners, so I would surely be in good company! Rather than trying to distract themselves, these folks will focus on their breathing and the feelings that they have in their muscles, closely monitoring each and every sensation as they move forward. The sports psychologist told us that this kind of focus can help runners relax. Let's see . . . what was I feeling? The hot sun blasting down on my head, neck, and shoulders. The straps of my heavy day pack—weighed down with my useless iPod—pulling down on my shoulders. One foot moving in front of the other. Each step more difficult than the last as my calf muscles were tested, forced to push and pull the rest of my body through the sandy terrain. *Hmm*, I thought, *I don't feel more relaxed. I just feel the pain with every step*. I dropped back behind Andry and went back to dissociation. I decided to focus on matching Andry step for step. He and I were close to the same pace, and I could see that he was suffering the heat as well. Watching him made me feel less alone in this. *Left. Right*. I chanted, feeling like a soldier.

And then I heard it. First soft and then louder. Something new to focus on. To help me dissociate. It was the whistling beats of our cook, Fidèle. Fidèle, like most Malagasy, was just over five feet tall—short and slim. He wore a colorful brimless cylindrical hat, a T-shirt, and shorts. While Andry and I hiked with a labored gait, Fidèle walked with a skip in his step, and as he walked, he whistled. He was pretty good at it too. I chuckled to myself: Who needs an iPod when you have a whistling chef?

CHAPTER 9

The Trouble with Zebu

June 3, 2006

> *For the hike into your field site it is a good idea to carry some hiking sandals in addition to your hiking shoes/boots, in case you need to cross any bodies of water. It is also smart to carry extra socks for that same reason.*

"Grab my hand," Andry said and reached out to me.

We were crossing a bend in the Mahavavy River. The water came up to just below my hips and the current moved swiftly. I took Andry's hand and edged forward on my tiptoes to try and keep my day pack, which held my binoculars, camera, and money—everything that mattered—from getting wet. In front of me, I could see the others forging ahead through the water. Gendarme Andry had taken the lead, searching for the shallowest and safest way across—the path of least resistance. Andry and I were keeping step—in this together—fighting the current as we neared the shore. I could feel large rocks at the bottom of the river, pressing against the soles of my hiking sandals. The slick, curved formations made it all the more difficult to remain balanced, and a few times I had to stop, recalibrate my footing, and start again. No one had warned me about this part. But I had to keep moving. Before we began to cross the river, through the treacherous waters, the guards had stopped, a look of concern on their faces. One of them asked Sahoby a question.

"They want to know if you need them to carry you across," he translated.

"No!" I exclaimed, perhaps a little too forcefully. I caught myself and took it down a notch. I gave it another go: "No, thank you. I can do it."

I knew I could handle it, and I was bent on proving it. *Left. Right. Stop. Steady. Repeat.* I could feel the men's eyes on me as I navigated the water, watching me with grave concern—the only woman on the team. I was determined to show that I could handle the hike into Kasijy as well as any of the men. Before long, I reached the shore, where I found Shawn tying up the laces of his hiking boots.

"We'll need to switch back to boots for the next little while," he said without looking up. Then he stood, turned to the porters who were waiting behind him, and said: "*Alefa.*" Let's go.

Just like that? No break? I thought again about how tough the terrain had been so far. Even Yoho, as challenging as it had been, had not prepared me for this. Here, it wasn't about elevation. It was that heat. That sand. And now we had to forge through rivers? Each step sapped more energy than the one before. If I had known this, I thought, I would have done more training back in Toronto. I took a moment to switch my Teva sandals for my hiking boots, tying the laces a little slower than usual. I needed time to catch my breath. I could see Andry out of the corner of my eye doing the same. We exchanged a knowing glance.

Prospère was standing beside the two of us, waiting for us to ready ourselves. He didn't need to change his shoes—he only had the one pair. And as I now examined his choice of footwear more closely, I became concerned. Andry, Sahoby, Shawn, and I all wore sturdy hiking shoes or boots. Shawn had outfitted Andry and Sahoby with boots from Canada—high-grade Merrells. But the porters, the gendarmes, and Prospère had chosen different footwear. They wore a shoe that I hadn't seen since I was a child in the eighties—a shoe that I never thought I would see again, let alone in the remote villages of Madagascar. They wore jelly sandals! All the rage in the 1980s, they were close-toed and made of PVC plastic. I'd owned more than a few pairs back in the day, ranging in color from pink to purple to glitter-blue. The Malagasy men had opted for plain, clear plastic—nothing fancy. I had visited the Bata Shoe Museum back in Toronto, where I learned that the jelly shoe originated in the late 1950s and early 1960s, when plastic started to grow in popularity and designers began experimenting with it.[1]

How jelly shoes made their way to Madagascar I do not know. But now, as Andry and I struggled to tie our laces after crossing the Mahavavy, I could see the utility of the jelly shoe. Where I had to stop and start, changing from my Tevas when we crossed the river, and then back into my hiking boots to walk through the sand, the jelly-shoe guys could just forge ahead. Maybe they were on to something. Now I find myself waiting for the day when Mountain Equipment Co-op starts stocking a Solomon-brand jelly sandal.

We resumed hiking and I watched as the landscape changed. Tall trees sprouted on the slopes above the riverbed. This was a different forest than I remembered from Belize. There, I also worked along a river—Monkey River—but the forest was a lowland, semi-evergreen, broadleaf forest.[2] The climate there is tropical, with two distinct seasons: a rainy season and a strong dry season.[3] Kasijy's is a dry deciduous forest, characteristic of forests in western Madagascar, where periods of heavy rainfall alternate with long dry stretches.[4] Deciduous means that the trees shed their leaves annually, and now, with June being the height of dry season, we saw few leaves.

A lack of leaves could make things tricky, I knew, when the need arose to identify tree species. Like any behavioral ecologist, I was interested in understanding the evolutionary basis for animal behavior as it relates to ecological pressures. In other words, how does an animal interact with its environment in order to survive and reproduce? We explore how animals find sufficient food, avoid predators, and respond to environmental pressures, such as seasonal food shortages.

For a primate behavioral ecologist who studies arboreal (tree-dwelling), folivorous (leaf-eating), or frugivorous (fruit-eating) species, the ability to identify tree and plant species at a research site is one of the most important skills to develop. Just like humans, nonhuman primates constantly interact with their environment, moving through the canopy and consuming plant and tree leaves, fruits, flowers, and even bark.

When we were trying to find the howler monkeys in Belize, for example, knowing which tree species were fruiting or flowering and where to find them could help us make an educated guess as to where the monkeys would be. Different types of plant species and different parts of those plant species (i.e., fruit, flowers, buds, leaves) can also have varying levels of nutritional quality for the primates that consume them, which is of interest to many researchers. In Belize, we would constantly quiz each other on the way to and from our trail system.

"What species is that?" Travis would say.

I would stop, look up, and pull out my binoculars. Often, the easiest way to tell was by the shape, size, and color of fruit or flowers.

"A kaway?" I would suggest. We would learn the colloquial names for the trees and later have samples identified in a laboratory to match them up to their scientific genus and species name.

"Let's go check." Travis would lead the way to the base of the tree. For a kaway, one way to be certain was to make a small cut to the trunk of the tree. Kaway trees would bleed red sap.

Learning, identifying, tagging, and measuring trees was one of my favorite parts of fieldwork. It was a nice change from chasing monkeys, who could sometimes be hard to find and follow in the thick canopy. The trees weren't going anywhere. I looked forward to days when we would collect information on the phenology (the stage of fruiting and flowering) of the trees. We would also spend days measuring locations, tree height, and trunk size, and would record the

species names of all the trees at the site to better understand the forest composition. I loved those days because I knew that we would fill pages and pages of our
data books and we would leave the forest with a sense of accomplishment. That
wasn't always the case when we were pursuing those tricky howlers. Now, as we
hiked over the sand in Madagascar, I knew I had my work cut out for me: none
of the trees looked familiar.

It was important that I learn about the tree species, though, if I was going
to understand the ecology and ranging patterns of the different lemur species
we would observe in Kasijy. Research has shown that many lemurs are impacted
by the seasonality of food resources. For example, in the eastern rainforest, the
rufous brown lemur, which we would also encounter in Kasijy, has been known
to eat more leaves (a lower-quality resource) in times where fruits are scarce.[5] I
knew that the dry western forests, like the ones in Kasijy, experienced dramatic
seasons, with a dry season that lasted about six to eight months longer than those
in eastern rainforests.[6] It was probable, I knew, that we might see the lemurs
shift their behavior or ecology in response to the seasons. I kept an eye on the
canopy as we hiked.

We were trying to stay together as a group—people and carts—but we
soon discovered that zebu carts, though useful because they can carry a lot of
equipment, were excruciatingly slow. We hikers quickly overtook the carts, still
slogging through the sand and the heat. I continued to put one foot in front of
the other and trudged along. At about 2:30 p.m., we stopped in a shady spot
near the river for lunch and to allow the carts to catch up. In a bid to save time,
Shawn instructed Fidèle to cook only rice—not beans—and to serve each person
a can of sardines in oil.

"We want to get back on the trail quickly—we can't wait for beans to cook.
I want to get in as many kilometers as possible before it gets dark."

Fidèle doled out the rice and then, diligently following Shawn's instructions, handed each of us a red, metallic, rectangular can of sardines in oil. I
have to admit that this *may* have been the first time I had ever eaten sardines
out of a can. Certainly, it was the first time in memory. Don't get me wrong, I
am no stranger to meat from a can. My mom used to make us tuna and salmon
sandwiches all the time growing up. In Belize, we would often eat canned ham
(it tasted shockingly similar to bacon after three months living in the field!) or
canned corned beef. It's just that sardines had never been on the menu.

After taking it from Fidèle, I carried my bowl of rice and can of sardines
over to where Shawn was sitting. I sat down next to Shawn and carefully placed
my bowl of rice on the ground in front of me. As I peeled back the lid of the can
of sardines I wasn't disgusted or concerned—I am not a picky eater. But I was,
let's say, interested. What was I in for here? The oil spilled out a little as I peeled

back the lid. Inside, I found four slimy headless fish. I looked over to see what Shawn had done. *Okay, then.* He had just turned and dumped the works—fish and oil—onto his rice. I did the same, put down the empty can, and dug in.

"What do you think?" Shawn asked.

My mouth was full, so I hesitated. Interestingly, the fish still had bones, but they were soft enough that I could chew and swallow them.

Shawn continued without waiting for my response: "I like bringing sardines because it is a good source of protein and fat. Plus, with the bones you get some calcium."

I nodded, my mouth full of oily, fishy rice. I remember thinking that the sardines were delicious. Maybe the best meal I had ever had. After that hike, I probably would have said the same about any food. As I polished off the last of my rice, I grabbed the hose of my camel pack for a quick sip of water to wash down my meal. But I was getting nothing.

I looked over at Shawn. "I'm out of water," I said.

"You have your water filter on you, right?"

I nodded.

Shawn motioned to where Fidèle sat with the pots and pans. "You can ask Fidèle for a bucket. Fill it up with water from the river and pump out of the bucket. I might pump another liter once you're done."

This was it. The first time my pump would truly be tested. *Let's see how it holds up*, I thought. I walked over to Fidèle and pointed to a purple bucket that I knew he had been using to gather water for cooking.

"Aza fady," *Excuse me*, I said as I pointed to the bucket again.

Fidèle understood and handed me the bucket, which I then carted over to the river. I found a spot where the water looked relatively clean and filled up the bucket. I carried the bucket back over to the shade where Shawn was now lying down, his hat over his eyes. I set the bucket down, unzipped my backpack and pulled out my filter and a one-liter Nalgene bottle, which I figured would be the easiest receptacle to pump into—the soft plastic of the camel pack bladder would be hard to manage while pumping.

I slowly and carefully followed the same steps that I had used back in Toronto, standing over my parents' sink. Step one: attach hoses. Black hose to the filter that would go into the dirty water in the bucket, clear hose affixed to the Nalgene bottle attachment. Step two: attach float to the dirty-water hose. I plopped the dirty-water hose into the bucket and started pumping. *Oof.* This was way harder than it had been with the clean tap water back home. I pumped a few times when suddenly . . . *POP!* The dirty water hose popped right off the pump. Hmm, that had not been an issue at home. I put the dirty water hose back onto the pump and tried again . . . *POP!*

Shawn lay next to me, his hat over his eyes. At the sound of that second pop, he peered at me from under his hat. "The water might have too much sediment," he offered. "It is probably jamming up your filter."

I peeked into the bucket. Sure enough, although the water had looked clean to me from the riverbank, I could see now that there was a lot of sand and sediment settling at the bottom.

"Try refilling the bucket with water from that rocky area." Shawn pointed to a spot on the river a bit farther away.

I carefully lay my filter and hoses on top of my day pack, hoisted myself up, grabbed the bucket, and fetched some more water. I rejoined Shawn and he sat up this time and took a peek at the fresh bucket of water.

"That looks better," he said.

Let's try this again. Hoses attached. Black hose to the filter that would go into the dirty water in the bucket, clear hose affixed to the Nalgene bottle attachment. Float attached to the dirty-water hose. Hose in bucket. Aaand . . . pump. This time, although it was still much harder than it had been when I practiced at home, I managed to pump a whole liter without losing my hoses. When I finished and handed Shawn the bucket, I commented: "Wow, it sure will take a lot of time and effort to make sure we have enough drinking water every day!"

"Yeah, especially with all the hiking we'll be doing," he agreed. "At least when we get to our camp, we'll be able to pump the water while we are sitting in our hammocks."

Pumping completed, and water replenished, we rested for another hour. I could see in Shawn's face that the rest was about twenty minutes too long. We hiked as far as we could before it got dark, which was around 6:30 p.m. We fell short of our target destination, and the zebu carts were still behind us, so we set up camp on the beach near the river.

As we assembled our tents, Shawn pointed out: "All of our food is on those zebu carts. If they don't get here soon, we won't have anything for dinner."

I sighed. No dinner? But I was so tired and hungry. Even though we had eaten lunch later than usual, the energy I had expended on the hike through the sand and heat had made me feel ravenous. And now that Shawn spoke words I had feared to hear, I felt the hunger pangs even more acutely. I couldn't remember whether I had slipped a protein bar in my backpack. As Shawn continued to clip in his tent poles, I discreetly walked over to my pack that was lying on the ground nearby. I opened the front pouch and felt around. *Yesss*, I thought as my fingers grazed the familiar aluminum wrapper of a peanut butter Cliff bar. *That'll do.*

Shawn, Andry, Sahoby, Prospère, and I all retired to our tents while the porters and gendarmes stayed by the fire talking. I lay on my back and unwrapped my Cliff bar. I stared up at my darkened tent fly, slowly savoring every last bite

The largest living lemur, the Indri. One of the many lemur species that compelled me to visit Madagascar. *Photo by Travis Steffens.*

Antananarivo, the capital city of Madagascar. The characteristic Renault taxis are visible on the street. *Photo by Travis Steffens.*

Coquerel's sifaka showing off the vertical-clinging and leaping mode of locomotion practiced by many lemur species. *Photo by Travis Steffens.*

My first real taste of Malagasy cuisine at a roadside *hotely* between Tana and Maevatanana. Clockwise left to right: me, Andry, Sahoby, and Prospère. *Photo by Shawn Lehman.*

Post-lunch break en route to Maevatanana. From right to left: Sahoby, MICET driver, Andry, Prospère, me, and our second MICET driver. *Photo by Shawn Lehman.*

Conquering the challenging task of getting the MICET van onto the ferry from Maevatanana to Mahazoma. *Photo by Shawn Lehman.*

Navigating the "road" between Mahazoma and Kandreho. *Photo by Shawn Lehman.*

Extricating the MICET truck from the muddy road between Mahazoma and Kandreho. *Photo by Shawn Lehman.*

Our zebu cart en route to Kasijy. *Photo by Keriann McGoogan.*

Homes in Antanabaribe. *Photo by Keriann McGoogan.*

The gray mouse lemur, the species that holds the honor of the first wild lemur species I laid eyes on. *Photo by Travis Steffens.*

The bamboo lemur. There was a chance we would see this species in Kasijy. *Photo by Travis Steffens.*

Milne-Edwards's sportive lemur. This nocturnal species is known to range in Kasijy. *Photo by Travis Steffens.*

The full field team upon arrival in Kasijy. Front row left to right: Sahoby, Noël, Andry, Fidèle, and Prospère. I stand at the back with Gendarme Andry and Jean-Paul nearby in their camouflage. The other men are our porters and trail cutters. *Photo by Shawn Lehman.*

Fidèle, our cook, preparing our lunch in Kasijy. *Photo by Shawn Lehman.*

The gendarmes on the bridge they built at our camp in Kasijy. *Photo by Shawn Lehman.*

The entrance of Ankarafantsika National Park—with Coquerel's sifaka— where I would later return for my PhD research. *Photo by Travis Steffens.*

A group of Coquerel's sifakas in Ankarafantsika National Park. This is the species that I would later return to study for my full PhD research. *Photo by Travis Steffens.*

of that 260-calorie bar. It was dark out, though it was still only 7:30 p.m. I wasn't ready to go to sleep, but I didn't have any of my books—they were all on the carts. After I finished my Cliff bar, I lay back again, still feeling hungry. I thought again about the sardines. Another can of those suckers would be perfect. And barbecue kettle potato chips would go well with the sardines. What I wouldn't give for a bag of those right now. I thought about my mom's macaroni casserole. My favorite meal growing up. I loved the way the noodles on the top got all crispy and brown. I thought about a chocolate cake I had baked back in February for my mom's birthday. That cake was *so good*. Dense and moist, with a buttercream frosting. Obsessed, or maybe possessed, I donned my headlamp and pulled out my journal—at least I had brought that with me. I began a new page. At the top of the page, I scrawled: *Things I will eat when I get home*. Without hesitation, I wrote:

Barbeque chips (kettle roasted)
Macaroni casserole
Homemade chocolate cake
Oreo blizzard from Dairy Queen
Movie popcorn—with layered butter
Eggs Benedict

I paused, my pen hovering above the page. *What am I doing?* I thought. *Why am I torturing myself like this?* I put down my journal and closed my eyes, willing myself to sleep. As visions of food swirled around in my head, my stomach still growling, I focused my mind on the porter's voices echoing outside my tent. I couldn't understand what the men were saying, but the musical tone of the Malagasy words they strung together was soothing. Eventually, I drifted off to sleep.

Next morning, the zebu carts still had not arrived, which meant no breakfast. That Cliff bar was the last of my snacks—the rest were on the cart with my other personal items. The hunger had officially set in, and my mind flashed to my list—what I wouldn't give for some eggs Benny right now! There was nothing to be done, and so we packed up and hiked the few kilometers to the agreed-upon meeting point and waited for the carts to catch up. They arrived at around 10:30 a.m. and Fidèle made breakfast: *vary maraina*.

As Fidèle handed us our bowls, Shawn asked, "Is this the first time you've had breakfast rice?"

I nodded and frowned at my bowl of steaming, wet rice. Sardines, I could handle. But this? This I wasn't so sure about. This dish is a typical Malagasy breakfast—something like oatmeal for the Scots. It involves cooking the rice with extra water. Then, rather than letting the water all boil off, they serve it wet. Simple. Not very flavorful, though, and definitely not high in nutritional value.

"Don't worry," Shawn said. "I asked Andry to buy tubs of milk powder. We are saving those for when we arrive in Kasijy. Add a scoop of milk powder, a few raisins, and a touch of cinnamon, and this tastes just like rice pudding!"

We finished eating and cleaning up at about noon. And so, on day two of our hike, we departed once again in the heat. At around 3:30 p.m. we arrived at a small village named Antanabaribe, where we had agreed to wait for the zebu carts. This village would be the last stop for those carts. After Antanabaribe, there was no passable road for the zebu, and so we would need to hire porters from the village. Antanabaribe would be my first remote Malagasy village. As we drew near, Shawn slowed to let me catch him.

"Just so you know, the more remote the village, the less likely it is that they have ever seen or interacted with a *vazaha* . . . and that can manifest in a couple of different ways."

Shawn described some of his own experiences working in remote Madagascar. In villages where they only occasionally see *vazaha*, he said, people are usually quite curious. "You'll probably notice that they stare. They aren't trying to be rude; they just don't see a lot of white people. Plus, our height is always interesting for them." Groups of children, Shawn said, would sometimes edge closer and closer to you, little by little, until one of them finally works up the nerve to touch you. Then they all scream and laugh and run away.

In the most remote places, where *vazaha* rarely visit, Shawn said that the people are sometimes fearful. Back in Canada I had read about that. In some of the most remote areas, the Malagasy believe that white foreigners are what they call *mpakafo*, or "heart-stealers."[7] According to their folklore, pale-faced monsters wander around at night and rip out people's hearts. Peter Tyson found clues to the roots of this folklore in the 1892 writings of English missionary E. O. McMahon, where he speaks of the slave trade:

> One African who is now free told us how, when the dhow they were on was becalmed near the Madagascar coast and [a European] man-of-war boat was in chase, the Arabs called up the strongest of the slaves to row the dhow, and to make them work harder told them that those in pursuit were cannibals, and were only chasing them to catch them for food.[8]

"I've seen grown adults look at me with sheer terror in their eyes," Shawn said. "Sometimes people will even run away, the belief runs so deep."

Some years later, when I returned to Madagascar to work with Travis on his PhD project, I would experience this myself more than once. The first time I will never forget. We were driving into a remote area to meet some of the local village presidents, Travis with the driver in the front seat. As we drove, we approached a woman walking in the middle of the sandy road, baskets of vegetables in hand.

She looked over her shoulder at our approaching vehicle and suddenly, without warning, dropped her belongings and sprinted away in front of us.

Our driver slowed down and yelled something in Malagasy. Our vehicle was too fast for her, and eventually she dove to the side of the road, like a baseball player sliding headfirst into home. When she looked up as we drove past, I could see in her eyes a look that I will never forget. No one had ever looked at me that way before. It was a look of pure terror. The driver explained that she had never seen white people before.

The second most memorable experience came when we hiked into one of the remote villages to meet with their local president. Our Malagasy student assistant, Mamy, had found the man and introduced us. The president had traveled around Madagascar and so was familiar with *vazaha*. He agreed to speak with us about our research project and also to show us around the village. As we walked, we noticed that there was a strange lack of people around. At one point, Mamy started laughing and calling out something in Malagasy.

"What is it?" I asked.

Two women popped out of some bushes and stood meekly at the side of the trail.

"They were hiding because they are afraid of you," Mamy said. "They thought that you had kidnapped their president."

> When you arrive in a remote village, it is very important to do your best to adhere to local customs. Your student assistants can be very helpful in this regard. The Malagasy people are some of the most hospitable people that you will ever meet and in even the poorest and most remote villages they will usually offer food and sometimes even a place to sleep.

As we neared the village, we could hear pounding.

"What's that sound?" I asked Andry.

He paused to listen. "It sounds like they are pounding rice." Andry explained that rice must be pounded to separate the rice kernel from the shaft. In Madagascar, the women and children usually pound the rice by hand by thrusting solid wooden posts into large stone mortars. "The rice is more flavorful when it is pounded by hand."

The village of Antanabaribe, where we had arranged to wait for the zebu, was unlike any place I had ever seen. Surrounded by large trees and situated on a sandy and grassy clearing, it comprised about a dozen small houses made of wooden logs and red mud. All the houses had thatched grass rooftops and no chimneys.

Andry and Sahoby took the lead when we arrived, speaking with the handful of community members who were out and about. Shawn and I waited and listened as Andry asked for the community leader and explained who we were and what we were doing. The men and women politely nodded and smiled. Andry said that the community members were fine with us waiting there for the zebu carts and had agreed that we could set up our tents and spend the night. Soon, a man brought out a woven straw mat and laid it on the ground under the shade of the awnings of one of the houses. He gestured for us to sit down.

It was hot and so I was grateful to be in the shade. But I was tired and looking forward to the arrival of the carts so that I could set up my tent and lie down on my Therm-a-Rest mattress, this time with a good book. We waited. We chatted. We lay down and took a few catnaps. The villagers in the community also sat outside, chatting among themselves and periodically looking over at us—particularly at Shawn and me. Some of the younger children stared, curious about these strangely tall white people. Although Shawn had warned me about the staring, it produced a strange feeling inside me. I have never liked standing out from the crowd or being the center of attention. I was struggling. I could feel the eyes on me. All my reading material was on the carts, so I couldn't hide behind my book. I knew that the villagers sitting across from us were just curious. Shawn had warned me—they may not have seen a white woman before. I glanced over and saw them whispering and felt a knot growing in my stomach. It was going to be hard to get used to the prying eyes. I sighed. We had more pressing matters to be concerned about now. It was five p.m. Darkness was falling. Where were the zebu carts?

Shawn turned to Andry and Sahoby. "It looks like the zebu carts aren't coming," he said. "We need to make a plan. Our tents and all of our food are on those carts."

The two assistants swung into action, speaking with some of the villagers, explaining our situation. When they returned, they told us that the villagers confirmed that the zebu carts probably weren't coming. By now, the zebu would need to rest. And since it was getting dark, the carts and drivers would have stopped somewhere for the night.

My heart sank. What would we do? I didn't know if I could go another night without dinner. Andry must have noticed the look of panic on my face.

"The village will make us some rice and chicken," he said.

I felt very conflicted about taking the food—especially the chicken. Sure, I was hungry, but as I looked around the village, I saw only a few chickens scratching around and I suspected that these people didn't have too many to spare. I saw a dog lying on the ground, so thin that I could see its ribs. That dog hadn't moved in a long time either. *Was that dog alive?*

I couldn't help but think of my family dog, Cody, a copper-colored cocker spaniel crossed with a terrier, who my parents had adopted on a whim when I was fifteen. On the weekends, my dad and I would take Cody out for a long walk in Nose Hill Park, a natural environment park near our house in Calgary where we would let Cody off leash as we walked along the hiking trails. We had to drive to get there, and as soon as we put him in the car, Cody was beside himself with excitement. As soon as we entered the trail and let him off his leash he would take off, running at full speed. He would run and run until eventually he would glance back to realize that he was too far from us, his pack. My father would call out: "Cody!" and Cody would run back—at full speed—until he met up with us on the trail. No sooner would he get back than he would spot another dog in the distance and he would race full speed toward the dog, visit for a while, and run full speed back to us again. This cycle continued as we hiked until Cody was so exhausted that he would just lie down in the middle of the trail.

When we would get back home, Cody, normally so full of energy he was almost unmanageable, would sleep the rest of the afternoon away. I loved watching him sleep after those long walks—he looked so content. But the dog I was watching now, in Antanabaribe, did not seem content. In fact, I could barely tell if that dog was alive at all. And I knew, too, that dogs in Madagascar represented yet another conservation problem. Unlike back home, where my dog was pampered with two square meals and treats throughout the day, the dogs in Madagascar—even the dogs that aren't strays—are lucky to get any food at all.

Because of that, the dogs will roam the forest and eat what they can find. The problem is that what they find in the forests of Madagascar is endangered wildlife, like lemurs or fosa, found nowhere else in the world. And one of the main reasons why people own dogs in Madagascar is to aid them in hunting. One organization trying to address the issue of dogs in Madagascar was co-founded by a fellow University of Toronto alum and is called the Mad Dog Initiative. This group has been working in Madagascar since 2013 and is implementing a spay/neuter and vaccination program for domestic and feral dogs in villages surrounding Ranomafana and Andasibe National Park in Madagascar.[9]

But as I stood now in Antanabaribe, I noticed that the problem went way beyond the dog. I noticed that the villagers wore threadbare clothes and were themselves very thin, even by Malagasy standards. A few of the smaller children had distended bellies, a sign of severe malnutrition. The people here needed this food more than I did, that was certain. But Shawn had explained earlier that it would be rude not to accept an offer of food or other hospitality. "A few years ago, I was visiting a remote village and they wanted to offer me food, so they cooked some rice," Shawn said. "But they didn't have any chicken and so they added grubs to the rice. It was hard, but I had to eat it." In many places in

Madagascar, he said, to refuse hospitality from a stranger is considered *fady*, or a breaking of rules that govern situations and behaviors.

The simplest translation of *fady* is "taboo," though my Bradt travel guide notes that this definition doesn't begin to capture the nuances of the term. *Fady* can involve flouting beliefs related to actions, food, or days of the week when it is "dangerous" to do certain things. *Fady* varies from location to location, from family to family, and even on an individual basis.[10] The longer I spent in Madagascar, the better I understood that *fady* rules are varied and unpredictable. Bradt lists some examples:

> singing while you are eating
> handing an egg directly to another person
> asking for salt directly
> holding a funeral on a Tuesday[11]

The Malagasy use stories to reinforce the rules of *fady*, some of which benefit conservation. For example, the largest living lemur, the Indri, is named "Babakoto" in Malagasy. This name translates to "Ancestor of Man," and is born out of a legend of a young boy named Koto, who traveled in the eastern rain forests of Madagascar.[12] Koto climbed a tree to collect honey, but someone passing by cut the vines and he couldn't climb down. Koto fell asleep in the tree and was awakened by what he thought were the sounds of an evil spirit. He closed his eyes. But nothing happened. When he opened his eyes, he saw an Indri. The Indri carried Koto to safety. As a result of this legend, local people revere the Indri and it is *fady* to hunt these lemurs.

Sometimes pockets of forest are considered sacred and so it is *fady* to cut or burn in that area. However, sometimes tribes will find a loophole that enables them to break *fady*. For example, *fady* may prevent some people from hunting lemurs, but not from eating them. And so, they buy the meat from a neighboring tribe of hunters.

We politely ate our chicken. But the meal was only part of the problem— without my tent, where would I sleep? As though reading my mind, Andry informed us: "We can sleep in that house," he pointed to one of the mud huts. One of the villagers was going to let us stay in his house. He showed us to his home—our shelter for the night. There were just two rooms. In the main room, devoid of furniture, a woven straw mat covered the floor. That room was apparently used for cooking, as a small charcoal stove sat in the corner. In the back, a tiny closet-sized room held a single bed. Andry explained that the eight men (Shawn, Andry, Sahoby, Fidèle, Prospère, and the three gendarmes) would all sleep on the floor of the larger main room. I, the only woman, would sleep in the back on the bed.

I protested—why should I get the bed just because I was a woman? I asked Shawn if he wanted the bed. After all, he was a lot bigger than me. But Shawn wasn't having it: "You're not sleeping out here by yourself with a bunch of men we barely know."

As I lay down on the thin foam mattress to try and sleep, I was grateful to the villagers for opening their home. But I felt uncomfortable. Anxious. My feelings weren't to do with the poor mattress quality or the fact that I could hear the mosquitoes buzzing around my ears—I could cope with the physical discomfort. What was making me uncomfortable was the thought that I was taking this bed away from someone else. Whom did it belong to? Where was that person sleeping tonight? I was haunted by the thought that my being there meant someone who had so little had just given me so much—had given me a place to sleep, without even knowing me. In that moment, I knew that I had to give back to the welcoming Malagasy communities. I didn't know how exactly, but I knew that I would.

I had a fitful sleep that night in Antanabaribe. I woke up mosquito-bitten and mentally drained. But Shawn had spent the night on the ground with seven other men, so my sleep could not have been nearly as bad as his. At 7:30 a.m., as we filed out of the house, I hoped we would be met by the sight of the zebu carts. No such luck. And so, again, all we could do was wait. I wondered if the drivers had absconded with our equipment, but I kept that fear to myself. Finally, at nine a.m., the carts arrived. Andry, Sahoby, and Prospère went to talk with the drivers. I was keen to get to the cart where my personal items were stowed. I wanted to change my clothes and grab a few personal items. I watched as the drivers started to off-load the equipment, and I headed for the cart that held my tent and personal items.

"No!" It was the typically reserved Prospère who intercepted me with a very uncharacteristic outcry. "You cannot go over there."

"What? I just want to grab my bag."

"We can't touch any of that equipment," he said, earnestly. "That cart has fleas." He made a scratching gesture.

I sighed. You've got to be kidding me. I could feel my cheeks flushing and my blood pressure rising. Fleas? But with the way this trip had been going so far, I was not surprised. As we ate breakfast, Andry—likely noticing my dismay—assured me that if we left our equipment alone for a while, the fleas would go away. They were just attracted to the zebu, he said.

Now that our equipment had arrived—fleas and all—we could begin arranging for porters to help us trek the rest of the way to Kasijy. Andry had already determined that we did not have enough carriers, and the villagers had suggested that we use a combination of porters and boats to reach our destination. Trouble was, Andry and Sahoby couldn't find anyone who wanted to work with us. We pitched our tents. It was too early to sleep so I wrote in my journal:

> *Spent a HOT day in the village with nothing to do but dream about Kasijy. Will we ever get there? I sure hope that things will fall into place somehow, but right now the thirty kilometers to Kasijy appear insurmountable.*

The next day Shawn decided that we should move our tents out of the village to a camp near the river. The villagers had stayed up late, conversing in Malagasy, and both Shawn and I had difficulty sleeping. Not to mention that the heat didn't break until the early morning hours. A riverside camp would be cooler and quieter.

Although we were positioned outside of the main village, as the day went on, a group of young girls, some wearing bows in their hair, had gathered near our camp. They sat off to the side, watching our every move. I began to feel self-conscious and whispered to Shawn: "They sure are interested." Shawn observed that he and I might be the first foreigners these villagers had seen in their lives—that they were probably fascinated by how tall we were, by our tents, and our other unfamiliar equipment. Feeling like all eyes were on me, I glanced in the direction of the girls, who giggled in response.

Now that we had our supplies, Fidèle cooked us lunch—rice and beans, and after we had eaten, I noticed him scraping out the remains of the rice pot.

"What's Fidèle doing with that burnt rice?" I asked Andry.

"He's giving it to the village," Andry replied.

I felt guilty as I watched the girls gratefully accept the burnt dregs of leftover rice and carry it back to their families in the village. I was reminded again of just how much I had in comparison.

After lunch, Andry and Sahoby tried once more to arrange transport for our supplies, but to no avail. And so, by three p.m., we accepted the fact that we would spend yet another night in Antanabaribe. As I lay in my tent, I worried: How would we get our equipment and ourselves to Kasijy? In my journal that night I scrawled:

THIS IS TOO COMPLICATED!

CHAPTER 10

Crocodiles and Bandits

June 7, 2006

Four days had passed since we left Kandreho. Finally, on the fifth day, we made a breakthrough. Andry and Sahoby found some men with boats, and also a few more porters to help take our supplies as far as a village named Bemonto. The catch was that the men had only enough boats to transport half the supplies. So, we arranged that the boats would take one load, drop it off in Bemonto, and then come back to Antanabaribe and fetch the other half. Meanwhile, we would hike with the few porters we had left and wait for the boats in Bemonto.

In Bemonto, Andry and Sahoby also found two guides who would work with us in Kasijy through the duration of the project. Noël was a quiet, shy, older gentleman with graying hair. He exuded kindness and I liked him immediately. The other guide, Jesoa (meaning "Jesus" in Malagasy), was younger, maybe mid-twenties, and obviously a different sort. He was loud and brash, and I found him off-putting. I reminded myself not to rush to judgment. Jesoa might turn out to be a terrific guide once we got to the forest. In Belize, we had worked with many guides who came across as "pot-heads," but who in the forest were pros at spotting howler monkeys and identifying trees.

When the two boats arrived at our riverside camp, I could see why they could transport only half our supplies. These were no motorboats. These weren't even canoes. These were pirogues—small, dug-out, wooden boats. Our pirogue drivers—two teenaged boys—used wooden paddles and push poles for navigating the shallow waters.

After we loaded the boats with what supplies we could, we set off hiking to Bemonto. The path took us through sandy terrain along the river, and—no surprise—it was difficult. But I was so grateful to be moving that I barely noticed. In the early afternoon, we arrived in Bemonto, where, miraculously, the pirogues were waiting. Turned out the pirogues were far better than the dusty zebu carts. Why hadn't we found pirogues in the first place? Andry and Sahoby spoke with the village leaders and arranged for us to set up camp near the river. Next, they went over to check in with the pirogue drivers. Maybe things were finally starting to go our way, I thought. But as the pirogue drivers began off-loading our necessities, Andry and Sahoby returned, their faces grim.

"What now?" Shawn asked, as the two somber students approached.

"They refuse to go back for the rest of the equipment," Sahoby said. "They will bring our equipment from here to Kasijy tomorrow, but they are too tired to go back to Antanabaribe."

We spent the next twenty minutes trying to negotiate with the pirogue drivers. But they had zero interest in returning for the remaining supplies, and no amount of money could convince them. To be fair, they were just teenagers, maybe seventeen or eighteen years old. And driving a pirogue upriver didn't look easy. But we still had a problem to solve—half of our equipment and food remained in Antanabaribe. And so, we switched tacks and worked instead on finding more porters to hike back for the supplies. We finally did find a few men who were willing to head back at 2:30 a.m. to bring the remaining supplies.

That night, after dinner, Shawn retreated into his tent, but I stayed out by the fire with Andry, Sahoby, and the rest of the guys. Fidèle was making us ranopango for the first time. Ranopango is an important part of Malagasy food culture and Sahoby had been talking about it since leaving Tana.

"If you like to drink tea, you will like ranopango," he had assured me. "I will ask our cook to make it for you."

I watched as Fidèle expertly cooked up the drink. On the fire, he had two pots—one in which he had cooked rice for our dinner, and the other filled with just plain water. Fidèle had already served our meals, and as usual there was a bit of rice left over—about an inch thick, stuck to the pot. He kept the pot on the fire and continued to heat that leftover rice. Just as we could smell it starting to burn, he poured the now boiling water from the other pot on top of the rice and let it absorb the flavors. He strained the hot rice tea into our mugs and distributed them among us.

It was a cool night, and the hot beverage was comforting. It reminded me of genmaicha, a tea I had often drunk at Japanese sushi restaurants, which was made from green tea and roasted brown rice. Like so much else, the ranopango drink signaled the importance of rice culture in Madagascar. The Malagasy have rice at every meal. There was the wet breakfast rice and mofo (rice donut balls), the mounds of rice served at lunch and dinner, and now the after-dinner aperitif: rice tea. Incredible.

We sat around the fire and sipped our ranopango. We talked about our challenges and frustrations. This trip was testing all of us and we hadn't yet made it to Kasijy. Gendarme Andry asked me about Canada. He was curious to know if it was cold. Where did I live? How many people lived in Toronto? Two-and-a-half million? What kind of music did I listen to? Like most Malagasy people, he hadn't traveled outside of Madagascar. Then he said something in Malagasy, and everyone laughed.

Andry translated, "He says you look like Celine Dion." Well, that was a new one—but I was ready to run with it. The Québécois chanteuse is beloved in Madagascar. I had been in the country only a short time and yet I had heard Celine's voice blaring out of car radios and from loudspeakers at many of the stores we had visited in Tana. I nodded my thanks for the compliment.

When you are looking for nocturnal lemurs, the easiest way to spot them is by shining your headlamps into the forest. You'll know you've found a lemur when you see two glowing orange eyes looking back out at you. The nocturnal lemurs all have a tapetum, a reflective layer of choroid in the eye, which causes them to shine in the dark.

"Keriann! Keriann! Come quick!" It was Prospère, returning to the campfire from the toilet. His call startled us. Andry, Sahoby, and I rushed over to see what was wrong.

"There's a lemur," he said, excitedly. On the hike earlier that day, I had told Prospère that I had yet to see a lemur in Madagascar and that I was anxious to see one. We followed Prospère to the edge of the sand and toward some small shrubs. He shone his headlamp.

"There," he pointed.

I crouched and followed the direction of his finger with my eyes.

"I don't see anything," I said.

Prospère moved out of the way so that I could stand where he had been. "There, you see?"

And then I saw. Yes, there it was. Two small orange eyes stared back at me. They belonged to my first Madagascar lemur. Based on its size and location, I figured it was a mouse lemur—probably *Microcebus murinus*, the gray mouse lemur, which we knew was found in this area. The aptly named mouse lemurs are tiny and rodent-like in their appearance—the smallest of all the lemurs.[1] I'd seen many a photo of these little guys cupped in the palm of someone's hand. They are so small, in fact, that the National Geographic Society website provides a graphic showing their size relative to a teacup (which they easily fit inside).[2] Looking at the gray-colored creature in front of me, it was hard to believe that it was a primate. It reminded me of E. B. White's famous anthropomorphic mouse, Stuart Little, but with huge eyes (good for seeing in the dark), a short snout, and protuberant ears. I managed to get a couple of good visuals of this little guy—I noted its primate features. Its grasping hands, and its long, thin tail.

While it might seem odd that I spotted my first lemur in a small shrub near a village, it wasn't all that surprising. In his prior research on edge effects, Shawn had determined that mouse lemurs are very tolerant of forest edges and disturbed habitats.[3] This species can live in secondary forests and degraded habitats. Mouse lemurs are agile and quick, known to move on all fours within and between trees and shrubs.[4] They consume a varied diet consisting of invertebrates (like beetles), and also fruit, flowers, nectar, and even some small reptiles and amphibians.[5] While the mouse lemurs target small prey for food, their size makes them a target for larger predators, primarily birds like the long-eared owl.[6] Gray mouse lemurs make use of tree holes or leaf nests to sleep, partly to avoid predators, like the owls, but also because these sites act as a thermal buffer against high temperatures.[7] Amazing—an animal the size of a mouse, but with the complexity of a primate.

As we stood in Bemonto, attempting to steady our lights on the small mouse lemur zipping around in the shrubs, I could see that our slow reactions and weak LED headlamps were no match for this quick-moving primate. Every now and then, though, I would catch a glimpse and feel a new rush of excitement. After just a few minutes, the lemur darted away and disappeared.

Two and a half weeks I had been in Madagascar, and finally I had seen my first lemur. In that moment, all the frustrations from the days of hiking, waiting, and negotiating melted away, and I remembered why I had come here. I was here to study primates. I was here to learn more about lemur behavior and conservation status, to contribute to the knowledge base for Madagascan lemurs—a contribution that I hoped would also help conserve the most endangered group of primates in the world. Conservation research, I knew, is much needed to help inform management of endangered habitats and wildlife. In fact, two Japanese researchers—Hideyuki Doi and Teruhiko Takahara—explored global patterns of conservation research and its impact on conservation. They found that there was a positive relationship between a country's proportion of published articles on conservation and ecology (relative to all scientific articles) and factors related to endangered species, including the total number of recorded endangered species and proportion of protected endangered species.[8] Research ensures that the countries that need more conservation attention are getting it. And, in turn, that research acts as a driver for conservation. In other words, my being there in Madagascar to study the lemurs *mattered*.

The next day was another hot one, and we were stuck in place yet again, this time in Bemonto village, while we waited for the porters to return with the other half of our equipment. My books, of course, were still in Antanabaribe. I vowed to stash one of my smaller paperbacks in my personal backpack when the gear did arrive. We sat around, talked, and built our names out of some large rocks next to the river.

I yearned to talk to Travis, if only for a few minutes. So much had happened—was happening—that I longed to share it all with him, and to hear what he had to say. Shawn let me send a quick message over the satellite phone, but what could be said in so few characters?

Miss you. We are on our way to Kasijy, but lots of delays. Hope you are well.

After lunch, Shawn announced that he was so hot that he was going to go for a dip in the river.

"Great idea," I said. A swim in the cool river sounded like the perfect response to the heat.

But there was a sudden and simultaneous outcry from our Malagasy companions: "No!" They explained that the Mahavavy River is home to Nile crocodiles, which are among the most dangerous of all crocodilians. Found in Madagascar rivers, freshwater marshes, and mangrove swamps, these reptiles are huge.[9] Males can reach up to six meters in length and weigh as much as 730 kilograms.[10] "They will eat you," Andry said. Human proximity to crocodile habitat in Madagascar makes for interactions and conflict. And Nile crocodiles, which eat mainly fish, are known to attack other animals, including humans. Our gendarme friend from Kandreho described a recent incident when a crocodile ate both a fisherman and his zebu.

I was no stranger to crocodiles. In Belize, I had surveyed the Bladen branch of the Monkey River for black howler monkeys. Once every two weeks, Travis and I would pay one of the locals, Darryl, to drive us to the mouth of the Bladen River where we would put in our two-man kayak and slowly paddle fifteen kilometers downriver, looking for monkeys as we went. The trip would take two days, and we would camp overnight on a sandbar next to the river. We paddled this stretch of river twelve times over the course of my six-month research project. The seventh time we went, Darryl asked us about crocodiles.

"You must see a lot," he remarked casually as we drove, listening to his reggaeton music.

Travis and I looked at each other, puzzled. We had heard that there were crocodiles in the river, but neither of us had yet laid eyes on one. In fact, while camping on a sandbar, we would often bathe in the river before heading to bed at night. Darryl couldn't believe it. The river is full of crocodiles, he said. American crocodiles are among the largest in the world: six-meter males can weigh up to 907 kilograms.[11]

On that seventh survey of the river, we kept our eyes peeled for crocs in the water and on the sandbanks. About an hour into the paddle, Travis exclaimed, "I think I see one!"

"Where?" I asked, scanning the water.

Travis pointed to what looked like a floating log. "There," he said. "Have a look with your binoculars."

I did. And there indeed it was—our first confirmed crocodile.

And then the floodgates opened. On that day of the Bladen survey, we spotted six more crocodiles—some on the river, others on shore. We realized that the crocodiles had been there during our previous trips, but our brains just hadn't imprinted the visual. Any bird-watcher, or primate researcher for that matter, will tell you that spotting animals involves practice and patience. It sometimes requires two or three times of homing in on a member of a particular species before the brain readily receives the visual message. And now, on the Bladen River, our brains got the message. Travis, especially, was spotting crocodiles left and right.

"There's another one," he'd say, laughing.

As we neared our usual sandbar campsite, Travis stopped paddling.

"Hang on," he said, quietly.

"What is it?"

And then I saw them—two humungous crocodiles resting on the sandbar, not one meter away from where we usually pitched our tent. As our kayak approached, they startled and plunged back into the river, making a huge splash. That night, we skipped the usual bath.

Now, at the Mahavavy River in Madagascar, Shawn listened politely to the protests from our Malagasy team members but pulled me aside.

"I don't doubt that there are crocodiles in the river. But I don't think they are as much of a threat to humans as the Malagasy would have us believe."

What was the risk? We talked it over. Nile crocodiles had been known to kill livestock and people. However, wildlife trade and uncontrolled hunting had drastically reduced their population.[12] Humans hunt crocodiles for their skins. Recent studies have shown that because of a few reported attacks people believe that crocodiles are numerous in Madagascar, but this is not the case.[13] Researchers estimate that the Mahavavy River is home to less than one crocodile per kilometer.[14] Despite the protests, Shawn was going to take a dip in the river. I decided to skip it.

While Shawn was swimming in the water, Andry approached me with yet another complication. One of the porters had returned, he said. The president of Antanabaribe wouldn't let the porters take the remaining equipment without one of our core team members there to confirm that they were permitted to do so. One of the gendarmes volunteered to hike back to the village to set things straight. And so, despite having left in the wee hours of the morning, the porters did not arrive back in Bemonto with the second half of our supplies until seven p.m.—far too late to start the hike into Kasijy. By then, the pirogues had already left for Kasijy with the first half of our supplies. We had no way to communicate with the boat drivers to let them know of our delay. We spent another night in Bemonto.

We had hoped to set out on the final leg of our hike into Kasijy bright and early, but we got off to our usual slow start. Again, we couldn't find enough porters to carry all our equipment. And so Andry and Sahoby spent a couple of hours negotiating for another boat. They secured one and at eleven a.m. we finally left for what we hoped would be our last day of hiking. The terrain was tough and sandy, and the porters were dragging, weighed down by the heavy bags of food and equipment. At 1:30 p.m. we stopped for lunch. Everyone was tired. The day was scorching hot and, though we expressed impatience, the porters refused to start hiking again until nearly four p.m. By six o'clock we were losing light. We needed to find a place to camp.

"We're only five kilometers from Kasijy now," Shawn said, holding up his GPS. "It's so frustrating that we just can't seem to get there."

We pitched our tents on a sandbar next to the river and proposed to leave the next morning at six a.m., although at this point none of us could swear that we would ever get to Kasijy. The temperature dropped overnight, probably because of our proximity to the river, and I was grateful that I had brought a warm sleeping bag.

I awoke to the sound of voices. Early morning. When I emerged from my tent, I found ten or twelve strange men—not our porters—standing with the gendarmes at our camp. Andry had just gotten up as well, so I went over to him.

"Who are they?" I whispered, a little concerned. Were these the bandits that the Malagasy officials had warned us about?

Andry said no. These men were searching for their stolen zebu and had stopped to see if *we* were the thieves. Having determined that we were not, they were asking if we had seen their zebu. That is what Andry told me, anyway. To myself, I said, yeah, right. If you and your gang were contemplating a robbery, you would tell that exact story. These guys were bandits and no mistake. Fortunately, after an extended conversation with our gendarmes, the men went on their way. I told Andry, "I don't think they liked the look of those AK-47s."

We downed some coffee—no time for a full breakfast, as Shawn made clear to the team—and were back on our way to Kasijy. The terrain shifted now from sandy riverbanks to a long series of ups and downs through head-high savannah grasses. We had to watch our footing because large boulders lay hidden in the grass.

Shawn looked around as we walked. "This is a prime area for wind erosion," he said. "It'll be interesting to see how far into the forest the wind is able to penetrate."

One of the biggest conservation concerns in Madagascar is soil erosion, which occurs largely due to deforestation.[15] Erosion here frequently occurs along hillsides and gullies, which means large amounts of sediment drain into streams and rivers.[16] In fact, there is a terrible "joke" among conservationists who work in Madagascar. What is Madagascar's biggest export? Answer: Topsoil spilling out into the Indian Ocean.

Some areas lose as much as 250 metric tons of soil each year.[17] In fact, the red soils that spill off the island make for a devastating satellite image—one that shows Madagascar as bleeding. This soil erosion creates large *lavakas*, or holes, which can sometimes measure several hundred meters in diameter. Conservationists refer to these as "denuded" habitats, meaning that the environment is stripped bare and inhospitable. Nothing grows in the lavakas, and so these areas are useless to both lemurs and humans. The savannah habitat leading up to the Kasijy forest wasn't too far gone. Maybe our research could help save this area?

Eventually, we reached the edge of the forest near the riverbed where the boats and drivers were waiting with our gear. As I surveyed the beautiful tall forest, I laughed out loud. We had finally made it—Kasijy. We had left Tana twelve days before, and the journey had taken every ounce of strength, patience, and determination that I could muster. But we were here now, and I would finally realize my dream of studying lemurs.

The catch? Shawn would have to leave in four days. After that, I would be the only *vazaha* for hundreds of kilometers—and the only female *vazaha* for even farther than that. I couldn't help wondering: Was I ready for this? Obviously, I had to be. But where was Jane Goodall when I needed her?

Part IV

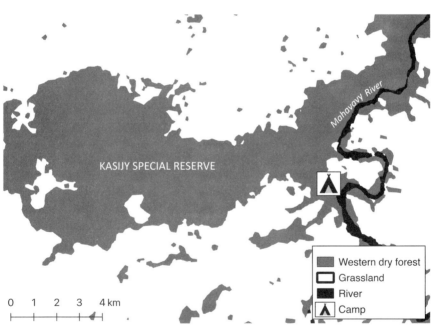

KASIJY SPECIAL RESERVE

Mahavavy River

	Western dry forest
	Grassland
	River
	Camp

0   1   2   3   4 km

Kasijy Special Reserve. *Map created by Travis Steffens.*
GIS data from J. Moat and P. Smith, Atlas of the Vegetation of
Madagascar / Atlas de la Végétation de Madagascar
(Richmond, UK: Kew Publishing, Royal Botanical Gardens, Kew, 2007).

CHAPTER 11

A Walk in the Park

June 9, 2006

As we entered the forest of Kasijy Special Reserve, I could hear a faint grunting sound. It sounded like pigs. I hesitated. In Belize, we had to be wary of white-lipped peccaries, a pig-like mammal that travels in large herds and has been known to attack humans. The locals told horror stories of naive tourists who had been ripped apart by groups of peccaries—if only they had known enough to climb a tree. Peccaries don't grunt like pigs, though—they bark and make chattering noises with their teeth. I had never seen a group of peccaries but had heard (and smelled) them several times while I was following the howler monkeys, at which point I carefully and quickly exited the area.

Now, I knew that noise didn't mean peccaries—they are not found in Madagascar—but I wondered if the grunters might be bushpigs. Scholars believe that the earliest settlers of Madagascar introduced bushpigs about two thousand years ago and that they later became naturalized. Bushpigs are large animals—the largest free-living mammal on Madagascar, in fact—with males weighing about seventy kilograms.[1] In his guide to the mammals of Madagascar, Nick Garbutt describes bushpigs as "a medium-sized wild pig with a shaggy appearance."[2] These creatures have been known to wreak havoc in agricultural areas, and so the Malagasy people consider them harmful and often hunt them. Although bushpigs are not dangerous to humans—they are probably more scared of us than we are of them—I still felt my heart leap in my chest. I couldn't shake the idea that all forest-dwelling pigs had a mean streak.

"What's that sound," I whispered, nervously, to Shawn.

"Keep walking—you'll see," he said, a knowing smile on his face.

We could hear the grunts getting closer as we followed our local guide, Noël, through a patch of bamboo. Sturdy stalks of green and yellow bamboo towered over us on all sides—stunning. Noël used his "coup coup," a sickle-blade attached to a long wooden handle, to cut a passable route through the bamboo.

The stalks were tough, so it was slow going. I had seen bamboo in Belize as well and knew from experience that it was one of the sturdiest plants around, which is why it is often used to produce furniture. When we emerged from the bamboo, we found ourselves in a small forest clearing, tall trees all around us. The grunting was louder now. I scanned the area, searching for the source.

"Look up," Shawn said, and pointed to the forest canopy.

The noise-makers weren't bushpigs. They were lemurs!

We were surrounded by a group of *Eulemur rufous*, rufous brown lemurs. My mind flashed to Shawn's lemur fact sheets. The lemurs in front of me were decidedly smaller than bushpigs—a solid medium-size relative to other lemurs (I remembered Shawn's laminated card: about two kilograms, with a length [head to tail] of 85–103 centimeters).[3] I remembered that the males and females differed in color (were sexually dichromatic), and I knew that the lemur in front of me must be male. I could have spotted him from a distance due to his dark, brick-red head and golden-red cheeks that contrasted against his olive-gray body.

Incredible. I knew that this species of lemur had not yet been studied in depth. According to my *Lemurs of Madagascar* field guide, studies from the 1970s that were originally thought to focus on this species are now known to have treated *Eulemur rufifrons*—a different species altogether.[4] The new understanding derives from studies of geographic ranges and genetics. The upshot is that we know very little about the conservation status of rufous brown lemurs, though the IUCN Red List of Threatened Species categorizes this species as "vulnerable," with habitat loss and hunting listed as the two biggest threats.[5]

I did know a little about the genus that rufous brown lemurs belong to—the *Eulemur*. They are a quadrupedal species that shows a cathemeral activity pattern, meaning that they are mainly active during the day, but are also active during the night.[6] Researchers attribute this cathemeral pattern to shifting seasons, diet, predator avoidance, and more recently to the level of human disturbance (through hunting, logging, or habitat fragmentation) in an area.[7] The lemurs have adapted a flexible behavioral pattern related to local conditions. I wondered what we would learn about the behavioral patterns of the rufous brown lemurs of Kasijy.

We stood watching the lemurs and tried to count them.

"I think there are six," I said, removing and unzipping my backpack to grab my binoculars.

Shawn scanned the area. "I think you're right. Looks like four females and two males. But wait—there's a seventh! Looks like a juvenile."

The *Eulemurs* show a varied social structure but are often found in larger multi-male/multi-female groups of sometimes up to twenty individuals.[8] As we stood there, the lemurs continued to grunt, and some of them moved lower in the trees, watching us with interest. These lemurs were curious. I adjusted my footing and a branch snapped under my hiking boot. But the lemurs didn't flee, as I had expected. In fact, they cocked their heads, like puppies awaiting instruction.

"Well, that's good news," Shawn said.

"What is?" I asked.

"These lemurs aren't afraid of us at all—they must not be hunted."

Second to habitat loss, according to Conservation International, hunting is the biggest threat faced by primates worldwide.[9] Across the globe, primates are hunted for food, to be used as bait, for medicine, and for their skins and other body parts.[10] Larger-bodied primate species are more frequently hunting targets—they provide more meat.[11] In the northwest, strong taboos prohibit hunting many lemur species, yet subsistence hunting continues within and outside protected areas.[12]

In a recent conservation strategy publication for the International Union for Conservation of Nature, Christoph Schwitzer and his colleagues reported that rural farmers in Madagascar often hunt lemurs using slingshots, spears, and traps, as a side activity to their agricultural activities.[13] Christopher Golden and his team have conducted extensive research on bushmeat hunting on the island. Golden points out that lemurs, like most primates, reproduce at a late age, have long periods of gestation, long life spans, and small litter sizes.[14] All this means that their populations cannot sustain even low levels of subsistence hunting. And the problem becomes even more complicated when you consider the needs of the Malagasy people. Schwitzer points out that Madagascar's growing population has increased sevenfold in the past one hundred years, and so pressure to provide food for families and a lack of sustainable livelihood strategies has created additional incentive for bushmeat hunting.[15]

If the lemurs of Kasijy Special Reserve were experiencing hunting pressures, we would expect them to flee at the sight of humans instead of coming down to investigate, as the rufous brown lemurs were doing now. Perhaps the lack of hunting was due to Kasijy's remote location or to local cultural norms and values. We could investigate the reasons later. But now, as I stood watching the lemurs watch me, I felt a flood of relief and excitement.

Shawn looked over at me, grinning. "Didn't I tell you that once we got here the lemurs would be dropping out of the trees?"

You made it! It's time to organize your camp. It is a good idea to have your guides build a few tables and platforms for storing food off the ground. Your cook should ensure that the food is organized and accessible.

The porters and boat drivers unloaded our equipment and supplies while Shawn, Andry, and I scoped out the area to find the best place to set up camp. We knew of one team of Malagasy researchers who had worked in Kasijy before us, and as we walked through the area we found evidence of their stay in the form of disassembled wooden sticks that looked as though they might have been used to build a table or two. As we walked, we got a whiff of a strong but appealing smell.

I sniffed. "Smells like . . . lemons?"

Sure enough, we found ourselves in a small patch of forest filled with lemon trees. The trees were small—they had slender trunks and stood only a little taller than me. But the crowns of the trees were filled with green leaves and yellow, plump lemons. Dozens of fallen lemons lay scattered on the ground.

Shawn laughed and looked around. "Wow, this is incredible."

We made our way through the lemon grove and emerged at a small stream. Surrounded by large rocks, the stream flowed into a series of shallow pools.

"Look at this," Shawn marveled. "This will be great for bathing or doing laundry." He turned to me, "Doesn't Kasijy feel like an oasis after our long struggle getting here?"

"Better believe it."

After some deliberation, we decided to build our camp in the clearing where we had first spotted the lemurs. That was close enough to the river that we could easily fetch water, and the secluded forest would provide shade and privacy. We also needed to be strategic about how far away we would be from our forest transects. Using a GPS and satellite maps, we could see how far we were from the forest edge, where we could begin cutting them. Each transect would run 1,250 meters perpendicular from the forest edge into the interior forest. From camp, we calculated that getting to the far end of any given transect would involve a three-kilometer hike. Not ideal. In the eastern forests, Shawn had been able to set up camp a mere 500 meters from the start of his transects. Here, we needed to stay close to our water source, and that meant accepting the longer hike.

Having chosen our location, we snapped into high gear, organizing our campsite. Our guides and the gendarmes collected the wood left behind by the previous researchers and began assembling a platform for our food and a table at which to eat (complete with a built-in bench). We had ready access to abundant water but getting to the river and back involved slogging through muddy terrain. As we now discovered, going to the river to wash was no problem, but returning through mud meant arriving back dirtier than when you started.

We asked our guides to build us some bridges. The guys also sorted out two toilets—one for the men and one for the only woman out there: me. Jean-Paul took great pleasure in showing me where my toilet was located, and proudly demonstrated (pants on, of course) the toilet seat that the guides had prepared for me. I have to say, I was relieved to have my own private bathroom.

For men, going to the bathroom in the bush is simple enough, but women must contend with certain practical matters. For one, we can't pee standing up, which means we need to account for things like wind direction and where we face relative to the slope of a hill. If a man needs to pee at night while camping in the field, he can just pop out of the tent, walk a short distance and he's away. Again, it is infinitely more complicated for a woman. In Belize, when we camped overnight on the sandbar by the river, I would often opt to just hold on through the night, rather than try to find a spot to squat in the dark. Later, some friends would awaken me to the joys of using a "pee bottle." This sounds gross, but it can be a savior if you're winter camping and would have been a godsend in Belize.

Aside from the daily challenge of where and how to pee, women also have to worry more about personal hygiene while out in the field. The last thing you want is a yeast infection or a urinary tract infection, but these things do happen, especially when you have no easy access to baths or showers after long days of hiking. This means careful selection of underwear material and bringing such extra supplies as unscented bathroom wipes.

Last but not least, women must contend with menstruation—another thing a male field scientist need not worry about. This complication presents itself at the packing stage. How many boxes of tampons or pads do you need for a three- to six-month stint of fieldwork? Can you resupply somewhere nearby or do you need to bring what you need for the whole time? These supplies take up space in your luggage. And what should you bring? Pads? Tampons? A mix? Oh, and make sure to pack extra Ziploc bags to pack out the used products if you will be going somewhere remote. Also true: a woman can find her menstrual cycle thrown way out of whack when she gets to the field. I know women whose periods became sporadic, or who had extended, month-long periods. Terrific. All this means you have to be ready for anything.

In 2016, Emily Graslie, a science educator at the Field Museum in Chicago, posted an amazing YouTube video on these matters.[16] She provided practical tips on how female scientists can handle their periods while doing fieldwork in remote locations. Her video opened a floodgate of discussion on the topic, which is not often addressed by academics. I found the open discussion of real, practical considerations empowering, and wasn't alone in this experience. Every female scientist faces the same challenges. Conservation biologist Anne Hilborn, famous for studying cheetahs, weighed in and linked to a Twitter discussion she'd had earlier that year.[17] My favorite line in the video comes near the end, when Graslie jokes, "As one researcher once told me, a little period leakage is nothing compared to the explosive diarrhea that keeps sending your lab mates sprinting to the woods every five minutes."[18] Gotta love it.

While the guys built our camp, I decided to establish my personal space. I scoped out a spot far removed from the others. *Perfect*, I thought. *This way, I can have some privacy.* I made to lay out my tent footprint, when Shawn approached.

"I would prefer that you find a spot a little closer to our tents," he said in a firm but gentle tone. "I don't want you to be too far away."

"I'll be fine," I said, shrugging off the suggestion. I didn't want to listen to the snoring from the gendarmes' tent.

"No, Keriann. I'm sorry to play the supervisor card, but you need to set up close to mine and Andry's tents. I don't like you off by yourself, at least not while the gendarmes are here. We don't know them."

I sighed. I didn't need it spelled out. I knew what he wasn't saying. That I was the only woman among a group of men—some of them strangers. He didn't want me to be sleeping in an isolated location where I would be vulnerable and where he wouldn't be able to hear someone approaching my tent. I think the incident with the drunken stranger outside the government building in Kandreho had left us both a little shaken. And then there was the incident with the AK-47. Although I didn't want to sleep near the group—I didn't feel I needed protection—I couldn't disagree with Shawn's logic. And so, I found another spot next to Shawn's tent and started setting up house.

I pitched my tent, strung a hammock between two sturdy-looking trees, and created a makeshift shower and change room, which consisted of two tarps strung up between a few trees and a solar camp shower—my solution to the privacy and hygiene issues. As I was putting the finishing touches on my shower, Andry came to fetch me.

"Noël is about to make an offering," he said.

"An offering?" I asked and followed Andry to the main area of our campsite.

We gathered round and quietly watched as Noël crouched next to the largest tree near our camp. He closed his eyes as he spoke in Malagasy, and he gingerly placed an offering of rice, sugar, and biscuits on the ground at the base of the tree. Andry explained quietly that Noël was offering the food to the Sakalava ancestors and asking their permission for our team to work in Kasijy. According to anthropologist Michael Lambek, Sakalava people are concerned with service to and care of royal ancestors, also termed *fanompoa*. In an effort to avoid giving offense, the Sakalava people will dedicate special ceremonies to their ancestors and their property (their bones, relics, tombs, houses, and artifacts).[19]

As I watched Noël conduct the ceremony, I found myself tearing up. I felt privileged to be a part of this ceremony. From an anthropological perspective, it was an amazing thing to witness—to share in the rich Sakalava culture. But I was emotional because the ceremony, for me, marked what I felt was the true starting point of my journey here in Madagascar. We had made it to Kasijy. *Kasijy*, where I would remain for the next few months. The journey had taken its toll on all of us, and nearly broken me. There were moments where I felt as though we would never make it. But now here we were. As I watched Noël crouched next to the base of the large tree, softly speaking his words—his call to his ancestors—I could feel the forest around me. I could feel the lemurs that lived there. The chameleons. The birds. Shawn and I exchanged glances. All the difficulties we had faced on the way no longer mattered. I knew that right now, right here in remotest Madagascar—this was precisely where I needed to be.

> *Once you have set up camp, it is time to cut your transects. You should arrange to hire three or four of your porters to stay at camp for a few days to work as trail cutters. Usually, it works well to have one man in front with the others following behind to cut, and with you bringing up the rear, using your compass to direct the lead cutter along a straight path. It is crucial to make as little impact as possible to the forest—make sure that you tell the team not to cut the trails too wide or destroy small trees.*

We awoke next morning to the grunting and chattering of lemurs. I bounded out of my tent (did I mention that I had become a morning person?) and greeted a rufous brown lemur group once again. Fabulous! The group of about eight individuals—a mix of males and females—was spread out across four different large trees in the camp. I looked up, unsure of where to focus my attention. Some of the lemurs were walking gracefully along the tree branches, using all fours to quickly and quietly move from branch to branch. Others appeared to be feeding—I could hear plant debris raining down to the ground from above. Some debris fell near where I stood. I stooped to pick it up. I held the small green, spherical fruit in my hand. I giggled as I turned it over and noted a bite missing. Lemur bite marks—amazing. I knew that the *Eulemur* primate group ate a diet of mostly fruit, but that the precise dietary composition varied depending on the season and location.[20] *Eulemur* populations from the west, where we were, have been shown to have a lower diversity in their diets than species in the north and east.[21] But no one had done an in-depth study of the rufous lemurs here in Kasijy. I wondered what our team might be able to contribute.

I heard the branches shake above me. I looked up and saw a female and a smaller infant jumping from one tree to another. The infant clung to the female's stomach as they moved, watching its surroundings. Female *Eulemur* are reported to carry their offspring ventrally and weaning occurs between six and seven months.[22] Females of this genus are not typically dominant. In fact, in some of the most egalitarian species, like the common brown lemur, females have priority over males when it comes to food, and they often lead group movements.[23]

As I watched those group movements, I realized that our camp must be smack-dab in the middle of this group's home range. Understanding primate home ranges—defined mainly by size, shape, degree of overlap, and location—is a key area of study among primate behavioral ecologists. In ecological terms, an animal's home range is simply the area where it spends most of its time, where it finds the resources it needs for survival (food) and reproduction (mates); it's usually measured over the span of one or several years.[24]

Ecologists are also interested in the animal's day range, or the area through which it moves in the course of one day.[25] The routes that a primate follows through its range can become predictable because they are often linked to the distribution of food resources, such as fruit and leaves, which remain stationary and turn up at fixed times of the year.[26] As the lemur group made its way through our camp, I felt keen to investigate our tree and plant situation. What was drawing the group here? Were those green fruits especially delicious? Or were we situated in the middle of a lemur "highway" of sorts, which passed through the delicious food trees?

We all gathered in the "kitchen" for breakfast. Fidèle had risen earlier than the rest of us. Shawn had outfitted him with a digital watch, and Andry had given him a quick tutorial on how to work it, helping set an alarm to alert him to begin preparing our meals. He had our breakfast rice and coffee waiting. Acting on Shawn's advice, I added some powdered milk, sugar, raisins, and cinnamon to my wet rice, and gave it a stir.

"How is it?" Shawn asked, as he approached our table.

"Wow, this *is* a lot better!"

Fidèle came by and placed a cup of hot coffee in front of me. Heaven. I took a sip but couldn't help making a face.

"Too sweet?" Shawn laughed.

That was an understatement. Fidèle had made the coffee over the fire "cowboy" style, by boiling the water, adding the coffee grounds to the pot, giving them a stir, and letting the coffee grounds sink to the bottom of the pot. He had then used a cloth filter to pour the coffee into our mugs. Trouble was, he had also added sugar directly to the pot of coffee. Fidèle asked if I wanted milk and I had replied "yes"—I typically drink my coffee at home with milk or cream, no sugar—and so he had added *sweetened* condensed milk to my cup. My Malagasy colleagues were now slurping their coffee, and some had even added another spoonful or two of sugar. I could only shake my head. When Andry sat down, I asked him to ask Fidèle to serve my coffee black next time, before adding sugar to the pot. That way, I could add a spoonful or two of milk powder.

"Tsy misy tsiramamy, *no sugar*," Andry told Fidèle and gestured to my cup.

Fidèle could not believe his ears. No sugar? How could anyone even consider drinking coffee without sugar?

As I discreetly poured my remaining coffee into our gray water hole, the four men we had hired to help cut transects arrived at camp. We welcomed them and Fidèle brought them each a plate of breakfast. After everybody had eaten, we began organizing to head into the forest.

"We'll start by cutting a small, passable route through the forest from here to the savanna," Shawn explained to me, Andry, and Sahoby. "Then from the savanna we will cut into the forest to create the transects. It's easier to ensure the transects run in straight perpendicular line if you start cutting from the edge inward."

As Andry and Sahoby were going over instructions with the men, Shawn noticed something.

"Where are their coup coups?" Shawn gestured to two of the men who stood empty-handed. The other two held the traditional Malagasy coup coup—the tool that Noël had used to cut through the bamboo when we arrived in Kasijy—which they would use to clear transect paths through the forest. Andry spoke to the men in Malagasy. Then he sighed: "They didn't bring them."

"What!" Shawn cried. "How can they help cut transects?"

And so, to cut the trails, we would have only two local men, plus Noël (who had wisely brought his coup coup with him). The other two men, Shawn decided, would stay behind with Jesoa to continue building and organizing our camp. The rest of us headed out. We planned to cut our transects on a north–south trajectory, perpendicular to the savannah edge. Noël took the lead, followed by the six of us—the two trail cutters, then Andry, Sahoby, Shawn, and me. Holding the GPS and the map, Shawn told Andry which way to go to reach the savannah. Andry then communicated to the men in front, who were making small cuts to lianas and brush to open the route.

This was not my first foray into trail cutting. In Belize, we had to cut and maintain trails through the forest. In Monkey River, there was a system of established trails, named with letters (e.g., "J" trail). These trails had been cut to follow the routes that the monkeys took, making it easier to keep up with them. Because Hurricane Iris had hit the field site shortly before we arrived, we faced a lot of thick, tangled undergrowth. From time to time, the monkeys would veer off into an area where we had no trails, which meant we sometimes lost them, especially if the area was tangled with vines and spikey plants. In that case, we would hire one or two local guys to help us clear a trail through the thick forest. Sometimes, these trails ended up feeling more like tunnels because they were covered over with messes of vines.

If we needed only a small, passable route to reach a tree we wanted to mea-sure, or if we wanted to clear existing trails, we would do that work ourselves. Upon arrival in Belize, one of the key pieces of field equipment we grad students would purchase was a machete. We would carry these long, sharp swords with us every day, clearing out nasty undergrowth as required. I had never felt so hardcore. How many women can say they've wielded a machete through the jungles of Belize?

Trail cutting was serious business and could be risky, because not every plant in Belize is "friendly." For example, the pokenoboy (*Bactris major*) is a small palm whose trunk, stem, and leaves are covered in needle-like stinging spines.[27] But there were other threats in the forest as well. Take the fire ants. Fire ants, one of the most common species of ant found in Belize, are aggressive. They are red, like fire, and tiny, only 2–25 mm.[28] I had several encounters with fire ants while clearing trails, as they do not like to be disturbed by machete-wielding researchers. Each encounter resulted in my having to strip off all my clothes in the middle of the forest to shake out the dozens of ants that had accumulated and made themselves at home inside my pants and shirt. Fire ants don't just bite. Anchoring themselves with their powerful jaws, they inject venom into their victim. And they don't just do this once—they sting multiple times, on and on.

But for trail cutters in Belize, the pokenoboy and the fire ants were relatively minor annoyances, really. The jungle also presented two life-threatening risks. The first was that, as a result of the hurricane, large, fallen trees remained sus-pended meters above us in tangles of vines. Cut the wrong vine and the entire mess would come crashing down.

Travis had a close call one day. While the two of us were working as research assistants for Kyle (the grad student with whom who we later hiked in Yoho), we needed to measure and tag all the trees in the field site, or at least as many as we could. We had set out one day—Travis; Kyle; Kyle's wife, Aliah; and me—to tag trees, and we had split up. The guys went ahead to clear paths, and Aliah and I followed, using the newly cut paths to reach the trees, measure, and tag them. Our work took longer because we had to wait for GPS satellites to mark locations, and so we fell a bit behind. We knew where the guys were because we could hear the rhythmic sound of their machetes chopping through the tangled vines. Suddenly, the chopping stopped, and we heard Kyle's voice ring out.

"Whoa, whoa, WHOA!"

Then came a crash—a large tree falling. Sudden silence. Aliah and I looked at each other. We dropped our equipment and ran to where the guys had been working. We found them standing on either side of a giant dead tree that had fallen between them after Travis cut the wrong piece of vine. Thankfully, every-one was fine. Kyle had noticed just in time. His yell had alerted Travis to what was happening, and they had both leaped out of the way.

The second deadly risk in Belize for trail cutters is the fer-de-lance snake (*Bothrops asper*), which the locals refer to as the "Tommy Goff." The fer-de-lance is a nocturnal venomous pit viper—a snake with a broad, flat head and a yellow, cream, or white-gray underbelly.[29] It is a large snake, weighing up to six kilograms and measuring up to 1.8 meters long.[30] The fer-de-lance is the most dangerous snake in Belize, and its fast-acting hemotoxic venom can kill.[31] Not only that, but the snake is known for being irritable, quick, and aggressive. Once startled, it has been known to bite repeatedly.[32]

We knew that the Monkey River forest was home to fer-de-lance snakes and so we remained on high alert. Researchers wore Kevlar knee-high snake-leggings and carried a snakebite kit. Our local friends had experienced more than a few run-ins with these Tommy Goffs, even finding them in and around their homes. One of the tour guides we knew from Ivan's, the local bar, had come home on his boat one night after taking some tourists out fishing. As he stepped, barefoot, off his boat and onto the sandy shore, he landed on a fer-de-lance. He was bitten multiple times on his lower leg. His family and friends rushed him to a snake doctor, a natural healer, and after a few horrific days he miraculously recovered—his leg still swollen, but on the mend. We figured that many of the bites he received must have been dry bites; otherwise, he might not have been so lucky. Thankfully, I never had a run-in with a fer-de-lance during my fieldwork, but the risk weighed heavy on our minds, especially when we cut trails through dark messes of undergrowth.

Now, in Madagascar, the forest felt relatively benign. Kasijy had not recently been hit by a devastating Category 4 hurricane, so there was not much deadfall and undergrowth. More importantly, Madagascar's forests contain few dangerous creatures, except for a few mildly poisonous spiders and toads. Before we left Toronto, Shawn had joked, "You're going to love Madagascar after working Belize, Keriann. You can practically skip through the forest without having to worry about anything dangerous." No snake leggings required—what freedom!

Geology and plate tectonics explain the dearth of dangerous creatures and the incredible levels of biodiversity. Between the Cambrian and the mid-Jurassic periods, Madagascar was part of the supercontinent, Gondowana, nestled between what today are Africa and Australia.[33] About 150 to 165 million years ago, through movements of the earth's crust, the island separated from Gondowana and began drifting toward its present location.[34]

Geologists once believed that the country's unique biology arose when Madagascar parted from Africa, bringing with it a sample of African fauna. The island became known as a "lost continent." However, we know now that Madagascar has been isolated where it sits today for the past ninety million years.[35] At least forty million years ago, the Mozambique Channel widened to a point at which no African animals could possibly drift over accidentally.[36] We are looking at a minimum of 400 kilometers. So, Africa's creatures, among them elephants, monkeys, lions, and, yes, poisonous snakes, never made the journey.

This long isolation of the island continent makes it hard to explain how, exactly, Madagascar's incredible biodiversity arose, and especially difficult to determine how lemurs reached the island. Lemurs, we know, are found only in Madagascar, but the continent of Africa contains lemur-like primates such as pottos and bush babies. Early researchers reasoned that these related creatures exist in both places because the lands were once attached. However, the picture grows muddy when we look to the fossil record and what we know about primate evolution. We don't see primates of comparable evolutionary level to lemurs—the "euprimates"—until the beginning of the Eocene, about fifty-five million years ago—millions of years *after* Madagascar separated from Africa.[37] The question has thus been haunting scientists for years: How did creatures that didn't exist when Madagascar separated from Africa get onto the island?

The best explanation we have so far is that lemurs reached Madagascar by "rafting," unwittingly drifting across the ocean from Africa on large clumps of floating vegetation that became separated during especially bad storms.[38] Picture this: a family of small primates—something like mouse lemurs, ancestors to modern lemurs—are holed up in a tree during a terrible storm. Suddenly, their home is ripped off the mainland. It becomes a sort of natural houseboat, carried out to sea through the Mozambique Channel. Days pass, and the primates find themselves washed ashore on Madagascar. Sounds far-fetched, I know. In fact, this explanation is referred to as the "sweepstakes model," a term originally coined by George G. Simpson.[39] The idea is that dispersal events, improbable within a limited time period, become more likely when you consider the vastness of geological time. After all, we are talking millions of years.

Besides the odds established by geological time, several tangible lines of evidence support this idea of lemur rafting. First, researchers have observed cases of such over-water dispersal before. For example, in 1998 Ellen Censky and colleagues witnessed green iguanas making their way from Guadeloupe to the islands of Barbuda and Anguilla in the Caribbean, thanks to favorable ocean currents and appropriate wind speeds. We are speaking of traveling hundreds of kilometers—270 and 150, respectively.[40] In the late Eocene period (about thirty million years ago), Madagascar was fifteen degrees farther south than it is now, and the prevailing winds and currents may have been more conducive to dispersal.[41]

Even then, the shortest distance between Africa and Madagascar would have been about 400 kilometers, so the "voyage" would have taken roughly ten days.[42] How could the primates survive the journey? What about food? What about water? Some researchers suggest that compression forces from India's collision with Asia caused uplift in the seafloor of the Mozambique Channel about forty-five to twenty-six million years ago, reducing the overall distance that the raft needed to travel.[43] Or, perhaps there were intermediate islands between Africa and Madagascar, and the lemurs island-hopped their way over.[44] Other researchers suggest that lemurs themselves offer some clues as to how they may have survived. There are several species that engage in *torpor*, which allows individuals to survive periods of reduced food availability.[45] Today, several species of dwarf lemurs, including the fat-tailed dwarf lemur, which we had a chance of seeing in Kasijy, engage in seasonal torpor.

Before you head to a new field site, it is a good idea to create several Google satellite maps for yourself. Overlay the latitude and longitude so that you can navigate using your GPS and make several versions with different levels of zoom. I like to laminate these maps before I go so that they survive field conditions.

As we walked, we discovered that the maps Shawn had created were less than ideal. Don't get me wrong; at least we had something, and the maps were good considering the resources Shawn had at his disposal. Fact remained: we were having trouble navigating to where we wanted to go. We could see the savannah on the satellite image—it was very clear—but what the image didn't reveal was everything in the landscape between here and there. It was tough to see exactly where we would encounter hills—and we encountered many. Also, some patches of our map were blocked with cloud cover. All this made for very slow going. The trail cutters went ahead, clearing a pathway, and Andry and I used bright pink flagging tape to mark the route for our return.

As I affixed pieces of flagging tape to tree branches, Shawn entertained me with a story. In the area where he used to work, Vohibola, he had flagged one transect with pink flagging tape. One day, he arrived at the transect to find the flagging tape completely gone along a fifty-meter stretch of trail. He wondered if maybe animals had taken it—rats for their nest, perhaps. But a few days later, while visiting one of the nearby villages, he discovered the truth. Six or seven young girls greeted him, and they were wearing bright pink flagging-tape bows in their hair. He said he gave the villagers their own roll of tape.

A few years after this first sortie, when I again visited Madagascar, I would learn that sometimes people in the small communities collect candy bar wrappers, discarded by tourists or researchers, and use them to decorate the walls of their houses. Such activities seem amusing to outsiders, but now, whenever I throw away a Kit Kat wrapper, I remember how privileged I am to regard that wrapper as garbage. Here in Toronto, where I write, I have covered my apartment walls with dramatic landscape paintings done by my mother, a talented showing artist. But when you think about it, that Kit Kat wrapper *is* quite colorful, so of course villagers who have so little would put them to use.

"Quiet for a second," Shawn called. "I think I hear something."

In the distance, I could hear the sound of crashing leaves. I would know that sound anywhere. Chasing howlers had taught me that I was hearing an arboreal primate moving through the leaf canopy. I scanned for lemurs. And then, like a vision, they emerged—the sifakas. This was the third lemur species I'd spotted in Madagascar. Just one hundred and eight more to go.

Sifakas ("shee-fak-ahh") are a member of the family Indriidae and the genus *Propithecus*. According to my lemur guide, the Indriidae are considered "the most unusual of all primate families."[46] Unlike the rufous brown lemur and the mouse lemur, which are arboreal quadrupeds (moving through the trees on all fours), sifakas are famous "vertical clingers and leapers." Vertical clinging and leaping is a mode of primate locomotion. Individuals keep the trunk of their body upright in an erect posture on vertical supports, such as a tree trunk, and uses specially adapted hind limbs, elongated and powerful, to propel themselves from one vertical support to another.

With this type of locomotion, lemurs can clear ten-meter gaps in the canopy with a single bound. Think Spider-Man! While such primates are extraordinarily graceful in the treetops, silently leaping from tree to tree like pinballs bouncing through an arcade game, on the ground they become more than a little awkward. Their long, bow-shaped hind limbs compel sifakas to jump in bipedal fashion, shuffling along with their arms above their heads in what has affectionately been nicknamed the "lemur dance." That dance is one of Madagascar's biggest tourist draws.

The sifakas we now saw leaping through the canopy were Von der Decken's sifaka, *Propithecus deckenii*. This species is snowy white in color (how do they stay so clean?) with a black face. They are about 92 to 110 centimeters (36 to 43 inches) long from head to tail and weigh about three to four and a half kilograms (six to ten pounds).[47] Van der Decken's sifaka is diurnal (active in the day) and lives in groups of six to ten.[48] Like many of the species we would see in Kasijy, this sifaka is not well studied. The species is categorized as "endangered," according to the IUCN Red List of Threatened Species, with a population reduction of more than 50 percent in the past half century.[49] As is the case for most lemurs, the main threat to this species' survival is habitat loss and fragmentation due to fire, agricultural activities, and logging for charcoal production.[50]

Shawn and I bounded off trail to get a closer look. We crashed through the undergrowth as quietly as we could, but we terrestrial bipeds were no match for sifakas on their home turf. In just two of their "steps," the sifakas had moved twenty meters by the time we reached the base of the tree in which we had spotted them.

Huffing and puffing, Shawn looked at me, "Wow. It'd be tough to try and study their behavior, huh?"

No doubt about that. But that didn't mean I didn't want to try. I had been working on trusting myself and owning my successes and failures. The challenges that I had faced mentally, physically, and emotionally to get to Kasijy had all made me stronger and helped me fight that imposter syndrome whenever it began to emerge. I was here for my PhD dissertation. I was here on my own. I wanted to study how forest edges impacted the behavior and ecology of sifakas. The expression on my face told Shawn that I was contemplating the magnitude of the challenge directly ahead.

"It'll be hard," he said, "but you should give it a shot while you're here. Go out with Andry and try collecting some behavioral samples."

"A walk in the park," I said. We both laughed. "I need to get a sense of what it will be like when I come back for my full project."

It took us four hours to reach the edge of the forest—less than three kilometers from camp! We had hoped that when we reached the savannah, we would find the terrain easier to navigate, but we found the opposite. The grasses grew as tall as me in some places, and they shielded large boulders that we had to climb over and around. Did I mention that I'm not the most coordinated human being? I tripped repeatedly and Shawn didn't do much better. *Boy*, I thought, *this will be dangerous for nocturnal surveys*. In addition to the difficult terrain, the savannah offered no tree cover, which meant we were exposed to the heat: 40°C (104°F). Suffice to say that, by the time we reached our destination, we were all exhausted. We paused to regroup.

"It's 11:30 now," Shawn said, looking at his watch. "I think we should head back. We're all tired, and assuming it takes us another few hours to get back, we will reach camp in time for a late lunch. We can talk about what to do for the transects after that." And so, on that first day out, we didn't even begin cutting transects.

Noël took the lead on the return. He said he had a shortcut back to camp. But the so-called shortcut wasn't short and took us through some especially difficult savannah terrain. I had just a few shortbread cookies for snacking, and I had nearly exhausted my water supply. I felt mentally and physically drained. We arrived back at camp around three p.m. Fidèle was waiting with lunch and wondering what had happened to us. He had readied the food for one p.m., when he had been told to expect us. As we ate our now-cold rice and beans, Shawn and I discussed what to do about the transects.

"I am thinking that we might as well capitalize on the trail that we cut today," Shawn said, "rather than starting a new trail entirely."

"That sounds good. In Belize, we used existing roads, trails, and rivers."

Shawn kept thinking out loud. "We'll need to cut our own trails here because the direction that the trails run in matters for research on edge effects. To document the edge signal for the variables that we are studying—temperature, wind speed, light penetration, and forest structure—we will need the transects to run as close to perpendicular as possible from the edge to the interior. Here that means north to south." He tapped his satellite map with the back of his pen and continued. "That way, we can measure each of our variables at marked distances from the edge. We'll then be able to see where the patterns level off—for example, at what distance from the edge does light penetration reach a plateau? That indicates the limit of the edge effects for that particular variable. Each variable that we measure will have a different impact on the forest and the distance each one penetrates the forest will vary—so the edge effect for light might reach farther into the forest than, say, wind speed. Get the idea?"

I nodded. "You know I do."

He continued: "We'll then be able to map those edge effects onto the distribution of lemurs along the transects. That's how we'll figure out how each different species reacts to the impact of edges. In the southeast, I found that some species were tolerant of edges while others avoided areas near edges."

"If a species avoids the edge," I said, just to set him at ease, "then the amount of useable habitat available to that species is reduced. So even though Kasijy is about 240 square kilometers, if the edge penetrates 500 meters all around, and a species avoids that edge, then actually the amount of usable habitat for that species is far less than what shows on that satellite map."

"Exactly," Shawn said. "So, we still need to cut some trails. But, looking at the map, the trail we cut today runs close enough to a north–south orientation, so we can use that for one of the transects at least. We just need to flag it fully. Catch is, I'm going to have to sit out tomorrow," Shawn gestured to his knee, "all that hiking has wreaked havoc on my old football injury, and I'll need to recover before the hike back out to Kandreho."

Whoa, this was an unhappy surprise. "You can take the team out," Shawn said. "Just go to the edge and start flagging it from there. Check around the area near the edge to see where you might cut the other three transects."

"When will you have to start back to Kandreho?"

"Unfortunately for me, in two days. I hope that if I rest tomorrow, I will be able to come out to the forest the day after, but we'll see."

For some reason, even knowing the original schedule, I had expected him to stay a bit longer. I could feel my stomach lurch at the very thought of being here without Shawn. The journey had been so difficult up to this point, and Shawn had been our key decision maker—the one we all turned to when things went south. And things had gone south a lot. I was struggling now to see myself in that role. All the tough calls would come my way. Even during my master's research, I hadn't been in it alone. I had always had Travis by my side. I wanted to cry out: *Wait! Don't you see? I have never done this before! Don't leave me now!*

Somehow, I managed to keep quiet, but after we ate, I asked Shawn if I could use the satellite phone to call Travis and my parents. I needed a pep talk after the tough day we had had, and the realization that Shawn would soon be leaving. I couldn't get hold of Travis, but I did manage a quick call with my folks.

"Are you healthy?" my mom asked. "Keeping well?"

"Yes, just tired," I said.

"I can hear it in your voice," she said. My mom always did have my number.

"Getting here was tough and setting things up is a bit of a struggle. Plus, Shawn is leaving in a couple of days."

"But you won't be there all alone," my mom reminded me. "You have this . . . Andy, was it?"

"Andry," I corrected her. "Yes, and Sahoby. That's true." All at once I felt a wash of relief. I wasn't here alone.

"Hang in there, Beavis," my dad said, claiming the role of Butthead. "Just a few weeks left."

This was one of our shared television references, Beavis and Butthead being two uneducated, barely literate cartoon characters who, lacking any scruples or sense of morality, precipitate one monumental disaster after another. After I hung up, feeling a little better, albeit homesick, the satellite phone beeped. A few messages had come through—one for Shawn and one for Sahoby, written in Malagasy. I left the phone on and out of the case and brought it back to camp, where I handed it to Shawn in his hammock.

"Thanks," he said, and read his message. "Can you grab Sahoby and let him know there's a message for him?"

I called Sahoby over and Shawn set up the phone for him to read his message. As he read the note, he gasped and put his hand to his mouth.

"What's wrong?" Shawn asked.

"It's my father," Sahoby said. "He's very sick."

> **Satellite text message received from Ken McGoogan**
> Sounds like things are looking up now that you have arrived in Kasijy. Don't forget to write in your journal!

CHAPTER 12

Dropping Like Flies

June 14, 2006

Sahoby needed to return to Tana to attend his sick father. And so, two days later, on June 14, I sat in our "kitchen," pushing my breakfast rice around in my bowl as I watched Shawn and Sahoby, together with Prospère and the two gendarmes, pack their personal belongings. Shawn sat in his hammock to organize his day pack. He was resting his knee before the long hike. I stood, handed my still half-full bowl of rice to Fidèle, and started to ready myself for the forest. As I shoved some cookies into my backpack, Shawn called me over.

"How are you feeling?" he asked as he fiddled with a stuck zipper on his pack.

"Great," I shrugged, fighting feelings of insecurity.

"Good," Shawn said, still focused on the zipper. "You're smart and capable, so you should have no problem at all. Plus, don't forget that you have Andry to fall back on. Make sure you lean on him if you need help navigating the local politics or dealing with language issues." Shawn paused for a moment. "It is too bad that Sahoby has to leave—I had hoped you would have a second person to talk with. But there's not much we can do about that."

Five of us were staying: Andry; Fidèle; our two guides, Noël and Jesoa; and me. Andry was the only one of the men who spoke English or French reasonably well, though Fidèle and I communicated using a combination of my Malagasy and his beginner-level French, which he had learned in his Kandreho grade school. Once Shawn was gone, I was officially in charge. Every decision would fall to me, and what I decided would shape the entire project. I kept it to myself, but I wasn't sure that I was up for this. I worried that the four men, all of them older than I was, would see right through me.

Frustratingly, I knew already that respect isn't a given for women. To earn a living during graduate school I was hired by the university as a teaching assistant for an introductory biological anthropology course. I was responsible for teaching thirty-student tutorial sessions where I would review key topics and run group activities. I recall standing in front of the group of students for the first few classes and feeling as though they would eat me alive. When the students—especially the males in the class—would ask questions, I could sense the skepticism in their voices and mannerisms. It was subtle, but it was there: *Will this young woman know the answer?* But I did. I did know the answer. It took a few classes for me to demonstrate just what I knew, and eventually guards went down and I could feel the shift. Don't get me wrong, I am not opposed to working to earn respect. What irks me, really, is that my fellow male TAs didn't have to work at it. All they had to do was show up. The respect was automatic. Now, of course, the guys could certainly *lose* respect along the way. But, wow, what I wouldn't give to start from the top. I wondered how it would go in Kasijy now that Shawn was leaving. How much effort would I have to expend earning the respect of my male team members?

I was fighting inadequacy and doubt, but I decided to get organized and focus on making a practical, step-by-step plan. That meant tackling the first obstacle—finishing the transects and commencing our surveys. It was time to collect data. I developed a three-day plan. When Shawn and the others left, we would head out into the forest with some trail cutters we had hired from Bemonto village to cut our second transect. The trail cutters would head home for the night and return the next day. We would split into teams. Noël and I would survey the second completed transect while Andry and Jesoa went with the trail cutters to cut the third transect. That would leave just one more transect to create, which we would do the following day. How hard could it be?

That first day—the day that Shawn would leave us—Andry and I would stick to the "program," as Andry called it. Breakfast over, I approached Andry as he readied himself for the forest.

"Let's say goodbye to everyone," I said, "and then we can head out." Andry translated for Noël, Jesoa, and the porters, and he and I wandered over to where Shawn was now kneeling and neatly folding up his tent fly.

"Okay, Shawn," I said. "We're heading out to the forest now."

Shawn stopped folding and stood up. He offered his hand, and I was reminded of that first day we had met. "Good luck," he said. "Keriann, Andry." We all shook hands. "I'll send you a text when we get to Kandreho to let you know about the hike out."

"Sounds good," I said. I gave a little wave to Sahoby and the gendarmes. I nodded to Fidèle—we would see him soon—and led Andry, Noël, and Jesoa into the forest. And I didn't look back.

That day in the forest we started by hiking out to the forest edge and into the savanna. I used the GPS to find the location for our next transect, which Shawn had marked for us. Off we went—cutting trail from the edge to the interior and flagging our second transect. This time, Andry and I took turns navigating with the compass while the guides and trail cutters followed our direction up ahead.

"Mankavanana." *Turn to the right*, Andry would say. And then: "Mankavia." *Turn to the left.* I was reminded of the RuPaul song, "Supermodel," and almost chimed in with a cheeky "Sashay, shanté," but figured no one else would get the joke. *Does* anyone else get the joke?

We were in a rhythm, in the zone. We managed to get our second transect cut and flagged before two p.m. I tied the final piece of flagging tape to a medium-sized tree trunk, and we rejoined our other trail, which would lead us back to camp. We had been walking for a while—RuPaul's "Supermodel" now running in an annoying loop in my head—when the men stopped, and I heard Andry say, "Salama."

Salama? Who could he be talking to?

I looked up to find two young women—maybe my age or a little younger—coming the other way along our trail. They both were dressed in worn cotton patterned dresses and wore flip-flop sandals on their feet. One of the girls had her hair neatly tied back in braids and the other had her shorter hair tucked behind her ears. The girls exchanged words with Andry and giggled in response to something Jesoa said. When they saw me, the girls looked at me with a curious expression and smiled. I smiled back, but wondered: *Where had these girls come from?*

We didn't stay long to chat, and as we parted ways, our team heading back to camp and the girls in the opposite direction, I caught up with Andry.

"Who were they? What are they doing out in the middle of the forest?"

Andry just shrugged and replied: "They're from a nearby village. They are out for a picnic."

A picnic? Incredible. Here I was, sweating profusely, huffing and puffing as we hiked through what in my mind was the remotest wilderness of one of the most exotic countries in the world—truly *the middle of nowhere*—and the young girls were . . . frolicking. *On a picnic.* I had to laugh. Maybe Kasijy wasn't as scary as I had built it up in my mind. If those two young girls could come out here alone for a romp through the forest, then surely I could spend my days collecting data. Maybe I could do this—I could succeed in my preliminary research without Shawn and without Sahoby.

Back at camp, which felt a bit empty now, we cleaned up and assembled for lunch. As Fidèle began dishing up the rice I looked around. We were missing the trail cutters.

"Andry," I said. "Where are the other two guys? Don't they want lunch?"

"I'm not sure," he said, scanning the surroundings. Which is what he said whenever he had no idea.

"Do you think they would have left and gone back to Bemonto?"

"It is possible," he nodded. Then: "Wait—there they are."

The two men were walking toward us. And they were carrying something I couldn't quite make out—something big.

"What's that they've got? Is that—? No. . . ."

As they got closer, and I could see clearly, my heart sank. It was a dead freshwater turtle. A turtle they had hunted. Hunted and killed on my watch. Then, slowly, I realized that this wasn't just any old turtle. It was a Madagascar big-headed turtle (*Erymnochelys madagascariensis*). The creature gets its name from the size of its head, which is so big that it cannot retract it all the way into its hard, dark brown shell. If I had identified the species correctly, I was dealing with a full-blown tragedy. The Madagascar big-headed turtle is one of the most endangered turtles in the world. Listed as "Critically Endangered" by the IUCN Red List of Threatened Species,[1] it is regarded as one of the top twenty-five endangered turtles by the Turtle Conservation Fund.[2] Local people commonly hunt this species for food, and in some areas, it has gone extinct.[3] I will never be absolutely certain it was a big-headed turtle that the men had killed, because as soon as they saw the look on my face, they turned and ran in the other direction.

The two must have known what I knew: poaching endangered species is illegal in national parks and protected reserves in Madagascar—including reserves like the one we were in. And yet, as I had just seen for myself, illegal poaching persists. Hunting is second only to habitat loss as one of the biggest threats to Madagascar's biodiversity.[4] The threat of overexploitation is especially devastating for species like turtles because they require a long recovery time once their population levels decrease.[5] Turtles have low annual reproduction rates and experience delayed maturity, which means they have a tough time bouncing back from a loss of adults in the population.[6]

Research indicates that people will exploit the most profitable species first, meaning they seek out larger species. Madagascar has already seen the extinction of two species of giant tortoises, *Aldabrachelys grandidieri* and *Aldabrachelys abrupta*, which were found in the highland region of the island, and filled the ecological niche of a large, herbivorous or plant-eating mammal.[7] Scientists believe that Madagascar once was a source of dispersal for giant tortoises to the western Indian Ocean islands.[8] Giant tortoises disappeared soon after the arrival of humans on the island, and probably hunting was responsible.[9] The giant lemurs that once roamed Madagascar met a similar fate. There used to be seventeen different species of giant lemur, but these have all gone extinct since humans arrived.[10]

Now, here at camp in Kasijy, what was I to do? What would Shawn have done? I knew that Shawn would have been as upset as I was in this moment. We were both conservationists. We had both come to Madagascar to try and save endangered wildlife. That this turtle had been killed in association with the project was absolutely devastating. I knew that I could not keep this quiet. I had to do something, say something. But, what? The men were gone. I could see Noël and Jesoa watching my reaction closely, wondering what I would do. I pulled Andry aside.

"This is serious," I said. "We cannot have people who work with us poaching animals."

Andry nodded, gravely.

"It's not acceptable," I said, "and it goes against everything we are here to do." I could feel my face flushing and my eyes began to water. "Please call Noël and Jesoa over here."

With Andry's help translating, I laid down the law. "Tell them that there is no hunting allowed. Tell them that we will not under any circumstances hire those two porters again. Noël or Jesoa should go back to the village to find some replacements for tomorrow."

The two guides nodded and agreed that Jesoa would make the trek to see about more men.

"Andry, please ask Jesoa to get the names of the two men who killed the turtle. We'll need to report this to the reserve officials when we get back to Kandreho."

That evening after dinner, I retreated into my tent. I couldn't shake my upset over the poaching incident. I was here to work on preserving Madagascar's biodiversity. Those men would not have killed that turtle if they hadn't been here, working for my project. What could I have done to prevent it from happening? Should I have laid down the ground rules from the beginning? Just when I had thought I had a handle on things, I felt the project spiraling out of control. Of course, it had been unwieldy from the beginning. But now, I couldn't look to Shawn. I was in charge. And there it was again: imposter syndrome. I pulled out my schedule and studied it. I needed to take this one day at a time.

Once you have your transects cut, it's time to start collecting data. Diurnal surveys should take place on a rotating basis between seven a.m. and eleven a.m. and between two p.m. and five p.m. Nocturnal surveys should take place on a rotating basis between seven p.m. and ten thirty p.m. and between four a.m. and six a.m. Varying the survey time will enable you to find all the different species of lemur that live in the area—both nocturnal and diurnal.

In the end, we would create only three of the four projected transects, compared with the six Shawn had set up for his projects in the southeast. Because of the incident with the turtle, we were short two men to help us cut trails. Jesoa had been unable to find anyone else to do the work. The forest in Kasijy also proved to be especially tough. Near the end of the second day, just as we were finishing up the third transect, the two remaining trail cutters abruptly stopped working. The men began fussing with their sickle-like blades—their coup coups. Growing concerned, I asked Andry: "What's going on?"

Andry shouted something out to the men in Malagasy. As they responded, he nodded with understanding and turned to me. "Their coup coups broke."

Crap. "What does that mean for us?"

"They cannot continue today."

"Okay." I thought for a moment. "The trails we have now are usable. But will they be able to fix the coup coups and come back tomorrow to cut the last transect?"

"I think so," Andry said.

The two men headed back to their village promising to return, though I knew better than to hold my breath.

The work was hard. We would walk for an hour just to get to the 1,250-meter mark, and then for another hour on the trails themselves. Our methods dictated that we should rotate the start location for each survey walk so that our results weren't biased according to the probability that particular species or groups ranged in the same area at certain times of day. This meant that on the days we started at zero-meters (at the edge of the forest), we needed a full two hours just to get to the starting point, and another hour to do the survey. And then there was the terrain.

"How are we going to navigate this at night?" I asked out loud as we picked our way up a steep hill covered in loose, jagged rocks.

Andry shrugged. "Very carefully?"

I laughed, then sighed, "Let's just hope we can get some good data out of all this."

Two days later, on our first official diurnal survey, four of us left Fidèle and headed for transect three. We all went together so that Andry and I could train the guides in methodology, including how to use the equipment. When we reached the 1,250-meter mark, we began the lesson.

My preliminary project comprised three components: finding lemurs (surveys); documenting abiotic (non-living) variables, including light penetration, wind speed, and temperature; and documenting biotic variables, including tree height and tree-diameter-at-breast-height. Today, we would focus on finding lemurs and documenting abiotic variables.

My master's project in Belize had also been all about primate surveys—looking for howler monkeys along roads, rivers, and trails to see how large the population was following a devastating hurricane. My supervisor, Mary, had come along for the first few weeks of my project to help me get set up and to train me on proper survey methodology. In Belize we had capitalized on existing waterways, trails, and roads so that we didn't have to cut new transects, which would have been costly and made more of an impact on the already-damaged forest. Our survey equipment was simple and low tech: waterproof "Rite-in-the-Rain" notebooks and pens, binoculars, and, the most high-tech item, a handheld GPS.

"When you walk along the trail, standard methodology dictates that you should walk at a speed of one kilometer an hour," Mary explained on our first survey walk. As it turns out, that is a very, very slow speed. We would frequently have to check our speed on the GPS and slow down. The idea is that the slower you go, the less likely you are to miss spotting a monkey. Keeping our speed in check was especially hard in the kayak because the river current wanted to move us along at a faster clip.

"If you see a monkey or a group of monkeys," Mary said, "you'll need to stop to write down all the information that you can about their location. Mark the location with the GPS and write down the number of individuals and their age-sex composition, their activity, what tree species they're in, how far they are from the trail, and anything else of note. At the end of your project, we'll be able to look at all of the sightings on a map and, using some basic math, we can extrapolate to estimate the population density of the entire watershed. We can compare that density to what we knew from before the hurricane to see just how much of the population we lost."

In the end, I found that the hurricane had reduced the population size by a whopping 88 percent. Out of that project, I became interested in how habitat disturbance can impact a primate species. In Belize, the disturbance had been from natural causes—the hurricane. In Madagascar the disturbance came at the hands of humans.

For our surveys in Madagascar, we would use the same standard methods I had used in Belize. But measuring edge effects required one extra set of operations. Following Shawn's previous research, we would keep track of temperature, light, and wind speed along the transects to see how far into the forest interior these variables penetrated. Along the transect, we had flagged various meter-marks, at increasing distances from the forest edge—zero, five, ten, fifteen, thirty, forty-five, sixty, and so on all the way to 1,250 meters—where we would measure light, temperature, and wind speed. We could only take these measurements after we had completed the lemur survey, so that we did not compromise the survey data. Along with my notebooks and binoculars, I also carried with me an Extech EasyView 30 Light Meter and a Kestrel 3000 Weather Tracker. This extra equipment made me feel hardcore—all my research up to this point had been low-tech and involved a lot of handwritten data sheets. Now, with my light meter and weather tracker, I felt like a real scientist.

We decided that we would survey transect three for lemurs starting at the 1,250-meter mark. Then, on the way back, we would collect our abiotic data. We slowly walked, single-file along the trail, all of us scanning the canopy for lemurs. About thirty minutes into the survey, Noël stopped and pointed.

"Gidro," he said, gently, in his dulcet tone. He looked at me and smiled. *Lemur.*

I couldn't help but grin from ear to ear when my eyes focused on the rufous brown lemurs that were feeding about ten meters away. My first "official" lemur sighting. One for the data books. I took down all the pertinent information—I could see six individuals. They were resting now, so I noted their behavior as inactive. Noël and I used the measuring tape to find the distance of the group from the trail, while Andry and Jesoa recorded the compass bearing. We were about 567 meters from the forest edge. Afterward, I slipped the logbook into my backpack. We continued our walk. And then . . .

"Gidro!" A more animated call came from the boisterous Jesoa. He pointed, and I followed his finger and saw a group of four snowy-white Decken's sifakas, all sitting together in what can only be described as a lemur gabfest. I turned to Andry.

"This might actually work."

He grinned.

After we had collected the necessary information—we were now about 417 meters from the edge—I paused to take one last look at the group. Kasijy, I knew, was one of just three national parks where this species was known to occur, and at the time Decken's sifaka had yet to be studied in the wild.[11] We did know that this species was vulnerable due to forest burning to make way for livestock, but that was about it.[12] There was so much to learn about this species.

Noël and Jesoa were terrific lemur-spotters, and the team was already getting the hang of the methods. We were on a roll—we could do this. When we reached the savannah, we stopped for a quick rest and a snack, and then pulled out the fancy equipment to collect the abiotic data on the walk back.

We took turns trying out the equipment. It was new to the guides and me. Andry gave us each a tutorial on how to use the Kestral and the light meter. Noël and Jesoa could both read and write—Shawn had made sure of that when we hired them. Not all of those who lived in the remote communities could do so, because schooling was hard to come by. As Andry demonstrated how to use the Kestral, I saw Noël's brow furrow in concentration, and I wondered: What did he make of all this? We were the first foreign researchers to come into the area. Did he understand what we were up to? What did he think of our weird gadgets and strange methods?

In 2014, 15 percent of Malagasy youth aged fifteen to twenty-four had no formal education at all, and 58 percent had not completed primary education.[13]

French colonists in Madagascar established the first formal system of public schools in the country—one system for the elite, reserved for children of French citizens, and another for the indigenous peoples, which offered little in the way of training.[14] After the Second World War, and then the country's achieving independence in 1960, the government reformed public schooling and gave the Malagasy more opportunities.[15]

Today, the official school system is divided into primary schooling (ages six to eleven), junior secondary schooling (ages twelve to fifteen), and senior secondary schooling (ages sixteen to eighteen). The University of Antananarivo, where Andry was a student, had arisen in 1961.[16] It offers undergraduate certificates/diplomas, bachelor's degrees, and doctorate degrees. Although public schooling is available in Madagascar, access is a problem, especially for rural communities. UNICEF reports that 14 percent of communities have no access to a primary school. Many Malagasy children are taught by poorly trained teachers and have no school supplies.[17]

Noël caught onto our methods very quickly, and soon he and Jesoa were working independently as a team. The speed at which the two of them understood the project and the methods demonstrated, once again, that the only difference between us was circumstance. I had been born in Canada, where I had access to public schools and a university system. They had grown up in remote Madagascar.

We made it back to camp by two p.m., which was becoming our regular lunchtime. I was tired but happy. The day had been a success. We had finished our first lemur survey, complete with two sightings. And we had gathered some abiotic data to boot.

After lunch, I settled into my hammock to pump some water and take a rest before our nocturnal survey. We would eat dinner, sleep for a few hours, and then leave camp at two a.m. After I finished filling up my Nalgene bottle with clean water, I pulled out a novel and let out a satisfied sigh. *This* is more like it, I thought. Just as I was about to crack open my book, I caught movement out of the corner of my eye. I hoisted myself out of my hammock and scanned the forest. About 300 meters away, I saw them: sifakas! I grabbed my binoculars—any primatologist worth her salt keeps her binos handy at all times—and, slowly and quietly, approached the group.

There were five of them, feeding on small, round, green fruits. They vocalized a little when they saw me—their characteristic sneeze-like "shee-*fok*" sound—but after a few minutes went back to feeding. I stood for half an hour, just watching them. Then I grabbed my notebook and tried out a focal animal sample, just to see. Focal animal sampling is a standard behavioral observational sampling method used in primatology. The legendary Jeanne Altmann, who spent years studying the social behavior of baboons, described the method in a seminal 1974 article.[18] In focal animal sampling, the researcher selects one individual in a primate group (yes, the "focal" animal) and observes it continuously over the course of a sample period (ten minutes is standard), recording all occurrences of behaviors and the start and stop time for each. Before a study commences, the researcher creates or selects an ethogram, a behavioral catalog in which all the different possible behaviors are listed and defined.[19] I zeroed in on an individual and I managed to keep an eye on my chosen one for the full ten minutes.

During that ten-minute sample, for the first time, I saw lemurs groom each other. As I watched, I suddenly understood how they manage to keep their snowy white coats so clean: their toothcombs, of course! The toothcomb is a characteristic feature of lemurs and lorises, and it distinguishes them from other primates. As a teaching assistant at the University of Toronto, I had taught my tutorial students all about toothcombs.

The order of primates is divided into two major suborders: the Strepsirrhini, which includes lemurs and lorises, and the Haplorhini, which includes tarsiers, monkeys, apes, and humans.[20] One of the traits that distinguish Strepsirhines from Haplorhines is the toothcomb.[21] Found on the lower jaw of lemurs and lorises, the toothcomb comprises six long, slender teeth—two pairs of projecting incisors and two narrow, elongated canines.[22] While researchers are still exploring the evolution and function of the toothcomb in lemurs, we do know that they use it primarily for grooming—literally, to comb through their fur—and feeding.[23] And now here they were using it, right in front of me!

After finishing my behavioral sample, I decided to see if I could find the other characteristic traits of lemurs. So far, I had seen two of them in the wild: the tapetum, or eye shine, that I had witnessed back in Bemonto village, and now the toothcomb. Now, I checked out the sifaka's ears. Strepsirhines are also notable for having mobile ears.[24] From about fifty meters away, I zeroed in on an individual with my binoculars and couldn't help but laugh out loud as I watched its ears twist around in all directions in almost robotic fashion. That's practically a super power, I thought. Finally, I focused on the noses. Strepsirhines have long snouts and what's called a rhinarium, a patch of moist naked skin similar to a dog's nose.[25] Of course, I couldn't tell if their noses were wet, but their noses did indeed resemble that of my dog, Cody.

After about twenty minutes, the sifakas began to head off. They had finished feeding and had other places to be. As I watched them disappear into the trees, I felt a sense of pride. It had been a rough go, but I could see the next few weeks falling into place. We had cut our trails. I had a schedule. Lemurs were everywhere. This project might just work after all. I decided that now was a good time to try phoning Travis.

"Hello?" When I heard his voice, I felt tears come to my eyes. We had not managed to connect for several weeks. I knew I was missing him, but I hadn't realized how much until I heard his familiar voice.

"I can't talk long," I said. "But I just wanted to let you know that all is well. I saw three different groups of lemurs today, one of them right from my hammock at camp!"

"Three groups! And hammock lemurs?" I heard envy in his voice. We talked for a few minutes, and then I let him go. We had to ration the minutes on the satellite phone for emergencies. This was shaping up to be the perfect day—data, lemurs, and Travis. Who could ask for anything more? Just as I was about to put the phone back into its case, a message came through from Shawn:

> FYI: Back in Tana. The hike out was only THREE days. There is a trail that we didn't know about. The porters demanded more money. DO NOT drive back to Tana at night.

Hmmm. This was interesting. A three-day hike was readily doable compared with the ten-day slog we had endured on the way in. I wanted to know more about this easy trail. Why hadn't we been told about it before? And what was this about driving at night? I guessed there was more to this story. I put the phone away and headed back to camp, where I told Andry what Shawn had said.

"Can you ask Fidèle and the guides about the trail he mentioned?"

Andry called the three men over and translated my question. Fidèle quickly nodded. Yes, he knew that there was a walking trail that provided a much shorter route into Kasijy than the way we had taken along the river.

"Why did we hike in along the sand?" I tried to mask my exasperation. Now I found out that our ten-day hike could have been both shorter and easier.

Fidèle explained that it was because we had said that we needed to stay with our supplies. That meant following the path for the zebu carts and boats. Well, with that I couldn't argue. We did specify that we needed to stay with the equipment, an instruction our local porters and guides had diligently followed. They must have thought we were downright crazy to want to take that longer route.

This was an important lesson for me and would inform all my future work in Madagascar. Malagasy culture is so deeply polite that often if you don't specifically ask, people will not offer up information, and they sure as heck won't tell you any bad news if they can avoid doing so. This trait can be problematic, as I now learned to my dismay. If we had known there was a better route, we surely would have taken it. But the guides and porters figured we didn't want to go that way. Crazy *vazaha*.

"Okay," I said to Andry, calmly. "Please tell them we want to take the fastest, easiest route on the way back."

After a little more back-and-forth between Andry and the three men, Jesoa turned and walked away.

"Where's he going?" I asked. "I have a few more things I want to discuss as a group."

"Jesoa is going to Bemonto," Andry said, "and won't be back until tomorrow night."

"What do you mean?" I asked. "We need him for the surveys. We have a nocturnal survey in a few hours, and tomorrow we were going to split up and do two diurnal surveys. We can't do two if he's not here."

"He says he needs cigarettes."

"Cigarettes?" I said. Now I was struggling to mask my exasperation. Noël looked down at his feet.

I couldn't shake the feeling that Jesoa didn't respect me, and I wondered if he would have treated Shawn or another man this way. It was hard to know. All I knew was that I wasn't pleased. Unfortunately, I didn't have much recourse. I watched, helplessly, as Jesoa grabbed his things and headed on his way. He would be gone for two days.

During your nocturnal surveys, it is a good idea to have at least three people. I've found that it works well to have one person focus on the right-hand side of the trail, one person focus on the left-hand side of the trail, and the third person checking both sides. That way, you will be more likely to spot the orange, glowing eyes of a nocturnal lemur with your headlamps.

The alarm on my digital Ironman watch startled me when it beeped at 1:45 a.m. I was in the middle of an especially good dream about home, featuring cold drinks, my mom's macaroni casserole, and vanilla milkshakes. I shimmied out of my sleeping bag and pulled on my field clothes, which I had laid out before going to bed. I grabbed my backpack, which contained my binoculars, water, and survey headlamp, and popped out of my tent. I could hear Andry and Noël getting ready a few meters away. I pulled on my LED headlamp, which I would use for the walk to the beginning of the transect. We would start this survey at 1,250 meters.

As I stood waiting for the guys to get ready—I am notoriously early for everything—I took a few bites of my Cliff bar. I wasn't feeling like myself and figured I must be hungry and tired. I followed Andry and Noël into the forest and through the lemon grove. Twenty-minutes into the hike, just as we started mounting the hill toward the transect, I felt a wash of dizziness.

"Hang on," I yelled to Andry.

He turned around. "Are you okay?"

I placed my hand on a nearby tree and leaned against it. "Not really," I said, after a beat. "I feel dizzy." The forest around me was spinning. I squatted down and put my head between my legs. I felt weak. *Why, oh why was this happening now? Why couldn't my body just cooperate and follow our "program"?* I took a small sip of water from my camel pack. The water didn't help. I still felt light-headed. I sat down on the forest floor. I was so frustrated that I could cry, but I held back the tears. After a few more minutes, Andry crouched down beside me.

"Should we go back?" Andry asked, concerned.

I couldn't believe this. My first nocturnal survey would be a bust. I could press on, I thought. But then I remembered the rough terrain that we had to navigate leading to our transects—the loose, jagged rocks, and the steep hills.

"Probably. I'm sorry."

Andry turned to Noël and translated, and we retraced our steps to camp. As we walked, more slowly now to accommodate my fragile state, Andry glanced at me every now and then, checking to see if I was okay. Eventually he said, "Fidèle is sick too."

"He is?" I asked. "I had no idea."

It is a good idea to bring a medical book with you into the field, in case you or your team gets sick. The book that I use is called "Where There Is No Doctor: A Village Health Care Handbook." It is a very practical manual that gives clear instructions on how to diagnose and treat common diseases in the tropics.

When we got back to camp after our failed nocturnal survey, I hit the hay and slept straight through until nine a.m. When I woke up, I felt better. I must have just been fatigued, I thought. And perhaps it was my low blood pressure rearing its head once again. I made a mental note to put some extra salt on my food. I decided, too, that it was best to take the day off and rest, just in case I had something more serious. Andry and Noël agreed to work in the afternoon. They would do a diurnal survey of transect three. As I approached our kitchen area for breakfast, I saw Fidèle sitting on a log, his head resting in his lap.

"Is he okay?" I asked Andry.

Andry shrugged. "I'm not sure. He says he feels very sick. A fever. Do you have any medicine you can give him?"

Shawn had warned me not to give out my medications too freely: "You need to keep your medications for yourself, in case you get sick. Plus, you aren't a doctor, and so you won't know what they are sick with, or what kind of medical care they need. I sometimes will give away some mild medications, like Tylenol, but just watch that you don't become the local pharmacist."

I looked again at Fidèle. He was shivering. I couldn't let him suffer. I went to fetch a couple of Tylenol. Our team seemed to be dropping like flies. I was feeling better after my day of rest, but now Fidèle was down and almost out, and Andry had started complaining about feeling ill. I hoped it was just a bug making its way through the group and not something more serious. I was glad that Noël, at least, was feeling well enough to work every day. He and I went out the next morning and surveyed transect two and found two groups of rufous brown lemurs.

That afternoon, Jesoa returned with the news that the trail cutters were not willing to return. The work was too hard. He had brought some porters carrying more rice, which Shawn had arranged, and a couple of guys who had come to pick up their pirogues, which they had left behind. The men hung around camp for the day with Fidèle and Andry, while Jesoa, Noël, and I did an afternoon survey of our farthest transect. The newcomers would stay the night, and head home in the morning.

When we got back to camp that evening, Andry was still sleeping in his tent. He came out for a quick dinner but didn't eat much and quickly retreated to his tent, complaining of chills and a fever. I gave him a few Tylenol in hopes of getting his fever down. I hoped that one more night's rest would help Andry feel better. He was my go-to guy, after all. He had even more experience than me with data collection methodology. He had worked with Shawn for years. Even more crucially, he was the only person who could translate between the guides and me.

That night, I was sound asleep when I heard a voice.

"Keriann?" It was faint at first, and then louder. "Keriann?"

Andry was sick. Very sick. And, it looked like malaria.

CHAPTER 13

Do Not Kick Away the Canoe

June 23, 2006

Andry pulled some pills out of the pocket on the side of his tent.

"What are those?" I asked.

"For malaria," he said. "I started taking them."

I sighed. Neither of us was a doctor. How could we be sure that Andry had malaria? What, exactly, were those random drugs he had taken? What if they had side effects that made him worse? I tried to hide my anxiety.

"I think I need a doctor," Andry said, his teeth chattering audibly.

But it was three a.m., late-night-forest dark, and if this wasn't the middle of nowhere, it was very near. We would have to wait until morning before doing anything. I did my best to make Andry comfortable. He asked me for Tylenol and I gave him two tablets. I made sure he had water. Then I left him to try and get some sleep, promising that we would figure things out in a few hours.

Back in my own tent, I strapped on my headlamp and poured over *Where There Is No Doctor*. The book reminded me that there are many different serious illnesses for which fever is a major symptom—not just malaria, but typhoid, typhus, hepatitis, and pneumonia. You have to differentiate in order to treat the patient properly. I flipped to the malaria section.

Malaria, I read, is an internal parasite in the blood, caused by a mosquito bite. The mosquito bites an infected person, sucking up the malaria parasites in their blood, and then injects the parasites into the next person it bites.[1] My mind flashed to Fidèle, who had been showing symptoms a few days earlier. Malaria, the book said, "begins with weakness, chills, and fever. Fever may come and go for a few days, with shivering (chills) as the temperature rises, and sweating as it falls. Then, fever may come for a few hours every second or third day. On other days, the person may feel more or less well."[2] That also sounded right. "If you suspect malaria or have repeated fevers, go to a health center for a blood test.

In areas where an especially dangerous type of malaria called falciparum occurs, seek treatment immediately."[3]

Falciparum? My travel doctor in Canada had warned me about that strain, which has a high death rate. With falciparum, the infected red blood cells can lead to microinfarctions (small areas of dead tissue due to lack of oxygen) in capillaries in the brain, liver, kidneys, lungs, and other organs.[4] Cerebral malaria, which finds microinfarctions occurring in the brain, can cause that organ to swell, leading to brain damage, a coma, and ultimately death.[5] *Plasmodium falciparum*, the species of parasite that can induce cerebral malaria, is a risk in Madagascar. My doctor had said that falciparum is the dominant type of malaria in Madagascar. And now I read a sentence that made my blood run cold: "IMPORTANT: Malaria can quickly kill persons who have not developed immunity. Children, and people who visit areas with malaria, must be treated immediately."[6]

I put the book down. The sun was coming up. The day was dawning. I could hear the porters, pirogue drivers, and our guides talking quietly by the fire. The porters and pirogue drivers would eat breakfast and then return to their villages. I knew that I would need to make some tough decisions in a few minutes. I left my tent and looked in on Andry . . . and found him worse than before. Fever, sweating, shivering, glassy eyes. I grabbed the satellite phone and headed out to make some calls.

Out on the sandbar, I tried Shawn first. I got hold of his wife, Christine, who told me that Shawn was still on his way home to Toronto. She was sympathetic, but she was 14,000 kilometers away.

"Not to worry, I said. "I know what I have to do."

Speaking of my situation made it all too real. I tried to call Travis but got no answer. He was probably in the forest himself. I could visualize Andry shivering. I hesitated even to think it, but I feared for his life. Somehow, I had to get him to a doctor.

Shawn had told me that, in case of emergency, I should call MICET. I did that now and got hold of Haza, the office administrator. I told her about Andry. "I'm not sure what to do," I said. "Can someone come and get us?"

Haza connected me with Benjy, MICET's lead driver. Benjy calmly explained what I already knew but was hoping might have miraculously changed: that a driver could come only as far as Kandreho. He could send someone today, he said, to meet us in three days in Kandreho and return us to Tana. But we would have to make our own way to Kandreho. He said he would need to know today, though, because Malagasy Independence Day was coming up in three days, on June 26, and the MICET offices would be closing, the drivers taking their vacations. I told him I would call him back.

I had been standing, but now I sat down. I was on my own. I gazed out over the forest canopy and the winding Mahavavy, took a deep breath, and then, methodically, ran through my options.

Option number one. We wait it out here and hope that Andry gets better. In that scenario, if he does get better, we could resume our work as planned, delayed by just a few days. But Andry could get worse, and then we would have an even harder time leaving because now he would be weakened. Also, left un- treated, he could die.

Option two. I send Andry out with one of the guides on the three-day hike to Kandreho, where there is a doctor. In that scenario, he could make it to the town, see the doctor, get some medicine, rest, and hike back to Kasijy. Maybe he is back by the beginning of July, giving us a few more weeks in the field. Alternatively, he could make it out to Kandreho only to find that the doctor there doesn't have the medications he needs, and so he is unable to come back, leaving me here alone for a month to finish the data collection and dismantle the camp. And then there is the scarier alternative. He is too ill to make the journey, complications arise during the hike, and . . . he could die.

Option three. We use the porters and pirogue drivers who happen to be at our camp to help us carry out our valuables and essentials. We all make it to Kandreho. We see the doctor there, give Andry a few days of rest, and then return to Kasijy. In that scenario, I can make sure that Andry is okay on the hike out and monitor his condition. We can take him to the doctor in Kandreho, and he might well get the medicine he needs. We can then come back and finish the field season. But as with scenario two, it is possible that we would get to Kan- dreho to find that the doctor there doesn't have the equipment or medicine to treat Andry properly. We would then have to arrange for the MICET trucks to take Andry to Tana, a three-day drive each way. Again, he could die. Meanwhile, I would have to return to Kasijy, without a translator, to salvage what I could of the field season. My head was spinning.

Option four. We evacuate the camp. I call MICET back and arrange for a truck to meet us in Kandreho in three days. We make use of the porters and pirogue drivers to help us carry out our valuables and essentials. We all make way to Kandreho. In Kandreho, if we can, we consult a doctor. Then, we hop in the MICET truck and head back to Tana, where I know Andry can receive effective treatment for malaria or otherwise. The downside here is that this would termi- nate the field season, and I would have nothing to show for it. The upside is that Andry would surely recover. In this scenario, I would have to coordinate a team of Malagasy who speak neither English nor French to carry out a delirious An- dry, along with as much of our expensive field equipment as humanly possible.

I ran through each scenario more than once. In every one of them, Andry's life was at risk. There was no way around it. That was just where we were. I knew that the porters and boat drivers were about to leave, so if we were going to evacuate, it had to be now. I also knew that there was a ticking clock on my ability to arrange for a MICET truck. Again, I needed to let them know today what we wanted to do. I felt the weight of the decision. For a moment, I was paralyzed by fear. I thought: *I can't do this.* Then I thought: *Oh, yes, you can. You can because you must.*

I picked up the satellite radio, made my way back to camp, and went to check on Andry. I squatted next to the door of his open tent. He looked worse than ever. He was pale and shivering, teeth chattering. He was clearly in no condition to hike, which meant that if we were going to evacuate him, we needed those boats.

"Andry," I said, as calmly as I could. "Do you think that you could hike out of here without me?"

"No, I don't think so."

Okay, I thought, *that means I am hiking out of here.* Next: "Do you think that the doctor in Kandreho—the one that you took Sahoby to when he was sick—would have the medicine that you need?"

"I don't know," he said, between shivers.

And so, glancing around the camp, I made the decision.

"Andry," I said. "I think we need to shut down the project."

We would pack up our essentials and the expensive field equipment and pay the boat drivers to take us to Bemonto. From there, we could get porters or zebu carts to help us the rest of the way to Kandreho, where I would arrange for MICET to meet us with a truck.

"Noël should stay behind," I said. "Clean up the camp, and send the items that we can't bring with us to Kandreho with the porters when they get back."

"What about all the food?" he asked. Obviously, he too had been sorting through options.

I thought for a moment. "We'll just have to give that to Noël and Jesoa to bring to their village. There's not much else we can do. We can't carry it out of here. We just need to take enough for the next few days."

Maybe Andry would get better, but maybe he wouldn't. As far as I could tell, his life was in the balance. And that was too big a gamble. Our location was just too remote. It offered zero contingency plans. At this point, while leaving was still an option, we couldn't risk anything but evacuation. The satellite phone was our only line of communication to the outside world. It was time to make the tough calls.

After I told Andry my decision, he seemed relieved. He weakly translated for Fidèle, who took charge of the others. I felt bad, because Andry should have been resting, but I still needed him to translate for me. I couldn't see any way around that. Charades just wouldn't communicate the gravity of our situation.

Through Andry, I laid out our new plan and gave the order to break camp. We were evacuating.

I headed to the sandbar, called MICET, and asked them to send a truck. When I got back, I saw the men dismantling the camp. I stood for a second and watched the demolition—our home in the woods slowly vanishing before my eyes, and with it, my project. Was I making the right decision? I couldn't help second-guessing myself. Then I noticed that our boat drivers were missing— along with the pirogues. We needed those boats! I raced to Andry's tent, where Fidèle and Jesoa were talking, to see if they knew anything. It turned out that the boat drivers had grabbed a couple of bags of equipment and set out.

"What! This whole plan hinges on having those boats." I asked Jesoa to run along the bank and catch them. I gestured emphatically: *Bring them back here.*

While we waited for Jesoa and the boats, we finished packing. A Malagasy proverb danced through my mind: "Do not kick away the canoe which helped you to cross the river." I thought: *Do not even let it out of your sight.* I looked over at Noël, who was tying up and organizing the food. When I saw his face, my heart ached. He looked so sad. I thought about what was happening from his perspective. He had just landed a job, complete with a place to stay and free food for four months. And now, suddenly, it was all being snatched away. He would return to his village with some food, but not nearly as much money as he had anticipated. Here in Madagascar, I knew, of the roughly sixteen million people who live in rural areas, some 85 percent live below the poverty line, often relying on tiny plots of farmland for survival.[7] I wished, in that moment, that I could do better by Noël.

We managed to get everything packed and sorted by 8:30 a.m., but saw no sign of the boat drivers. Every second, every minute, every hour that went by meant one more second, minute, and hour that we weren't seeing a doctor. I had hoped that we could make it at least to Bemonto that day, but now the odds seemed slim. I did my best to keep calm, but I was being tested. The boat drivers got back at eleven a.m.—far later than I had wanted to leave.

We had three boats. Each boat could carry only the driver plus one additional person. Andry would go in one boat, I would go in another, and the small amount of valuable gear that we could bring would go in the third. To my surprise, my pirogue driver would be Fidèle. Something must have happened that wasn't communicated to me—for some reason, the third boat driver had backed out. At that point, all I cared about was that we were finally on our way to Bemonto. We said goodbye to Noël and Jesoa, who would finish cleaning up the camp, and started our return journey to Kandreho.

While a pirogue ride along a river in Madagascar may sound appealing, even romantic, it was nothing like riding in a gondola along the Grand Canal in Venice. I had savored that experience as a seventeen-year-old on a school trip, and I can attest that the two experiences are nothing alike. In the gondola, I had sat with my girlfriends, talking and laughing, comfortably perched on a cushioned seat while the costumed gondola driver maneuvered our boat slowly and delicately through the still canal water. In the pirogue, I sat on the bottom, legs stretched out straight in front of me, trying my best not to rock this unstable wooden dugout. We teetered awkwardly as Fidèle stood behind me with a long wooden stick, giving his all to push us through the water against the powerful current. As we worked our way forward, I discerned that Fidèle had little experience as a pirogue driver. In places the river was very shallow, and we would get stuck. In brief, we made slow progress.

Up ahead, I could see Andry's boat, moving marginally faster—though these boats weren't designed for speed. Andry was lying down in the bottom of the craft, covered in a Malagasy lamba. Lambas are large, rectangular cloths, usually made of cotton, and are a traditional garment for men and women in Madagascar. Malagasy people often wear them tied around their bodies like Indian saris, but they also use lambas to carry their babies, as sheets or tablecloths, and even to bury their dead.

The lambas are usually colorful, with bold patterns framing the edges, and the middle offering images of things important to Malagasy people, such as cattle or food. Some lambas, called *lamba hoany*, also feature a Malagasy proverb or message important to the region. I hoped the lamba was keeping Andry warm. He looked so frail, lying in the boat. I felt helpless watching him from where I awkwardly sat, Fidèle standing behind me, struggling to move our boat forward. After battling the current for four hours, we had yet to reach Bemonto. By four p.m., the drivers were spent, and they began steering into shore. I had no choice but to set up camp for the night.

"Andry, we need to tell them that it is very important that we leave very early tomorrow morning, so that we can get you to the doctor."

Andry translated, and the men promised we would leave by one a.m., so we could make it to Kandreho in two more days. But, as I had learned on the way to Kasijy, these early-morning promises often didn't pan out. We ended up leaving around eight a.m. the next morning. I hoped that, in the slow-moving boats, we would make it to Bemonto for lunch, and could carry on from there. But noon, one p.m., two p.m., three p.m.—all came and went, and still we fought the current. Finally, at four p.m., we reached Bemonto village—still a two-day hike from the nearest doctor.

When we got to shore, I saw that Andry's condition had worsened. As I helped him with his tent, he became delirious.

Shivering, he asked: "Can I have some bug spray, Keriann?"

"Bug spray? Umm . . . sure." I reached into my backpack to pull out my bug spray.

Andry grabbed the bottle and started spraying it all around his tent.

"What are you doing?" I asked, concerned.

"We must get rid of all the mosquitos," he said.

I gently took the bottle away from him, explaining that it wasn't that kind of bug spray. This was meant for a person's skin, not a tent. At that point, Andry started rattling on about the bad smell from the spray. He was feverish and talking nonsense, so I gave him some more Tylenol and helped him lie down in his tent. Now I was really frightened. Andry was delirious. I was scared to think about what that might mean as far as his condition went. I knew that delirium is indicative of at the very least an extremely bad fever—we're talking 104° Fahrenheit and above. I hoped that was all it was and that I had done the right thing by giving him the Tylenol. After I left Andry, I knew that I needed to communicate the gravity of the situation to the boat drivers, who were laughing and chatting by the fire. But I was in a predicament now because asking Andry to translate was no longer an option. He was too far gone.

I knew, too, that there was just no way that we could reach Kandreho in another day, or even two—not using these boats and traveling against the current. At this rate, the journey would take us another three or four days. Again, I called MICET on the satellite phone. I asked for advice. They said the driver was on his way. Short of that, there was nothing they could do from Tana. I told them we would likely arrive in Kandreho a day later than anticipated. I hung up the satellite phone, acutely aware that I was on my own.

We needed to get to the doctor as quickly as possible. That meant ditching the boats and hiking. We needed porters. When I got back to the tents, Andry was asleep. I found Fidèle by the fire. He spoke a bit of French. And, using a combination of his French, my rudimentary Malagasy, and some complicated gestures, I enlisted his aid in changing tactics. From here in Bemonto, we needed to hire some men to work as porters. We needed them to carry our gear and Andry to Kandreho.

Under Fidèle's direction, several men began building a makeshift stretcher. They found two sturdy pieces of wood and strung a few lambas between them, making a kind of hammock. Fidèle got in to test it out. It didn't look comfortable, but Andry couldn't walk. It would have to do. Two of the porters would carry Andry in that stretcher all the way to Kandreho, a distance of about thirty kilometers.

I managed to rally the team by seven a.m. the next morning. I had wanted to leave earlier, but nobody but me seemed especially worried about Andry's health. For the people of Bemonto, Andry's condition probably seemed innocuous. Probably they had seen worse. People in Madagascar die needless deaths from not just malaria, but such everyday ailments as diarrhea and respiratory infections.

While in Canada we have the luxury of vaccinations, children in Madagascar receive no immunization against polio, tetanus, or measles. Access to clean, safe drinking water is not guaranteed, and sanitation in rural communities is poor. I thought of the children we had seen with distended bellies in the communities we had passed through on our way to Kasijy—a sign of malnutrition. And so, although I recognized our situation as urgent, the men around me probably thought I was overreacting. They dealt with medical emergencies on a weekly or even daily basis. You win some, you lose some. This reality contrasts starkly with that we find in most Western countries. In Canada, we have universal health care and on short notice can usually find a good doctor, receive a diagnosis, and acquire any necessary medications.

Andry clutched his water bottle and tumbled into the makeshift hammock-stretcher. Two men hoisted the wooden poles onto their shoulders. I followed behind. We had also picked up another woman in Bemonto, who would hike with us to Kandreho. She was pregnant, so maybe she too was off to see a doctor. I welcomed the sight of another female. We smiled at each other.

Hiking out proved faster than poling along in the boats, and I hoped we could reach Kandreho, and the doctor, that day. Fidèle seemed to think it possible. We stopped for lunch at around two p.m., and then resumed. But as on the way in, the hike out proved difficult. I marveled at how the men were able to carry Andry uphill, downhill, through sand and mud, and across the river. That day we hiked for twelve hours. Yet still we failed to reach Kandreho.

When we stopped walking at seven p.m. to set up camp by the river, Andry got out of the stretcher and stumbled a little. Although I didn't think it possible, he looked worse even than before. He was no longer hallucinating, but he had a fever and sweats and had stopped eating. I thought about calling my parents or Travis, but there was nothing anyone could do to help me. Why worry people who were far away? I would call after I got safely out.

I noticed that Andry was running low on water.

"Here," I said. "Give me your bottle, I can pump you some more from the river."

I pulled out my water pump, took Andry's bottle and my own, and squatted down by the flowing water. I told myself everything would be okay. As I pushed the lever on the pump back and forth, I felt an increasing resistance. And then . . . POP! The hose popped off. Too much silt in the water. The pump was jammed. I managed to put the pump back together, but this particular straw almost broke the camel's back. It took everything I had not to start crying.

The next morning, we left at 6:30 a.m. Exhausted, we all wanted desperately to reach Kandreho. The final stage of the hike was the hardest, with countless steep hills to climb up and down. The pregnant woman navigated most of the hills like a pro but did need to scoot down a few on her bum. As I watched the pregnant woman next to me navigate her way down a large boulder, I couldn't help but think about how surreal this whole situation was. I looked around and saw the pregnant stranger, two men carrying Andry in a makeshift hammock-stretcher, and Fidèle whistling away beside me. It felt like a really bad and bizarre dream.

And then the whistling stopped abruptly. "Keriann!" Fidèle turned to me and pointed.

There, in the distance, finally, I could see the large village of Kandreho. Tears came to my eyes once more and relief washed over me. Andry was going to be okay. And then a strange thought popped unbidden into my head. *Man, I would kill for a cold bottle of Coca-Cola.* Suddenly, a bottle of cold Coke was all I could think about. As we continued to put one foot in front of the other, I was consumed by the idea of the fizzy, cold, deliciously caffeinated beverage. *If we make it to Kandreho and everything is okay*, I thought, *I will buy a huge bottle of Coke and chug it.*

We reached Kandreho, and the house of the deputé, at eleven a.m. There we feasted our eyes on a most beautiful sight: a truck parked next to the house, featuring the MICET logo. As we approached, the driver emerged, a concerned look on his face.

"Keriann!" the driver practically shouted as he burst out of the truck. "I was worried. They told me you would arrive yesterday."

"The hike out was more complicated than I had hoped." I gestured toward Andry who had squatted down on the ground and put his head between his knees. "We need to get him to the doctor. Can you help me?"

The two of us got Andry up and put him in the backseat of the truck. We drove the short distance from the deputé's house to the doctor's office, a tiny concrete building in the center of town. The driver called out in Malagasy as we approached, and a diminutive woman emerged—the doctor.

We brought Andry in and sat him down in a wooden chair next to the wall.

"He has had a high fever and chills for a few days," I told the driver, who translated to the doctor. "I think he might have malaria, but don't know for sure."

The doctor turned to Andry and asked him a few questions in Malagasy. Andry responded, weakly. She took his temperature and his blood pressure. Then, she turned to her cabinets. She pulled out a package of pills and handed them to Andry as she softly spoke in Malagasy.

The driver turned to me, "She says that he must start taking these pills immediately, but his blood pressure is low."

"Okay," I said, digesting the news.

"She says you should take him back to Tana, so the doctors there can confirm her diagnosis and do a blood test."

At the back of my mind, I had been clinging to the hope that Andry might recover here in Kandreho. He might recover and rest a few days, and then we could return to the field site, rebuild our camp, and begin again. But I wasn't going to act against the advice of a doctor.

To the MICET driver, I said, "Let's pack up the truck and head to Tana."

When we emerged from the doctor's office, Fidèle and the porters were waiting. I paid the porters first, shook their hands, and thanked them profusely for their help.

"Misaotra be," I said. *Thank you very much.*

Then I turned to Fidèle—the whistling chef, whose positive attitude I had so appreciated, and who had been a great support throughout the evacuation.

I called the MICET driver over to help translate.

"Please tell Fidèle thank you so, so much for his help. We could not have made it here without him."

I counted out Fidèle's salary, added a few ariary and, with tears in my eyes, handed him the money. Fidèle thanked me and shook my hand, as if in congratulations. I watched as he walked away, a happy-go-lucky skip in his step.

I got the MICET driver to help me coordinate the delivery of the rest of Shawn's field supplies from Kasijy to the deputé's house in Kandreho. If Shawn wanted to come back here, his equipment—mostly just plastic dishes and buckets—would be waiting. We had managed to bring the important supplies with us: measuring tape, Kestral, light meter, GPS, books. And there was another important thing.

"Two of our porters killed a freshwater turtle," I told the driver. "That's poaching. We need to make a stop at the National Park's office to report it before we leave. Do you mind coming in with me to help translate?"

After I filed a report with the park official, who promised to investigate, we made arrangements to drive to Mahazoma—just outside of Maevatanana—where we had dined with the deputé and his family on our way in. We had learned our lesson on the way in and hired a local from Kandreho to ride along and help us negotiate the road. Andry was resting in the backseat of the vehicle as the rest of us loaded up and readied ourselves to go. He looked marginally better. The medicine he had received from the doctor was helping.

As we waited for the MICET driver to finish securing our luggage, I hurried into the local store.

"I just want to grab one thing for the road," I told the driver as he cinched the rope down taut. In the wooden building, I peeked over the counter to see into the refrigerator. There was, sadly, no Coca-Cola, so instead I purchased three one-and-a-half liter bottles of the Malagasy equivalent, *Classiko*—one for each of us. That would do for now. I ran back to the vehicle and took my seat in the front. I twisted the cap and chugged the soda in long, thirsty gulps.

Your guides and cook will likely want to take off a few days around June 26th for Independence Day. Independence Day in Madagascar is a big deal and a big party.

When we arrived in Mahazoma, after an uneventful drive from Kandreho (if only we had hired a guide on our way in!), we found a hotel, ate dinner, and went to bed. That night, I was awakened by hoots and hollers and the sound of . . . gunshots? No . . . firecrackers. Then I remembered: today was June 26—Madagascar Independence Day.

Madagascar's history had fascinated me long before I got here. I had been excited to think that I would be in the country during the holiday. I had imagined celebrating the holiday with my Malagasy teammates in the field, preferably with some Robert's chocolate around the fire, and lemurs chattering in the background. I had not anticipated lying in bed in a darkened hotel room in Mahazoma.

The next morning, we ate a quick breakfast and began the drive to Maevatanana. As before, we needed to cross the river on the ferry. But when we pulled up to the ferry dock and piled out of the vehicle, we found nobody there. And the ferry, we could see, was on the other side of the river. My MICET driver went to investigate, and after a few minutes came back to report that the ferry wasn't running because of the Independence Day holiday. For some reason, this final hiccup did not surprise me at all.

"We're stuck?" I asked and looked over at Andry who was still tired and pale.

The MICET driver said not to worry. He negotiated a boat trip across the river, convincing a bunch of guys hanging out by the shore to hop into the river and pull the ferry across by hand. We then drove the truck onto the ferry and the guys pulled us back across. Only in Madagascar!

In the eight hours that it took to navigate the winding highway back to Tana, I had plenty of time to think. The crisis resolved, the reality of my new situation hit me hard. I still had more than a month left in Madagascar. My flight wasn't scheduled to depart until August 9. Returning to Kasijy was no longer an option. So where did that leave me? Maybe I could do some work at another research site?

Obtaining research permits, I well knew, often took several months. And even if I could lay hands on one, where would I go? Then I would have only a few weeks to set up a site and collect data. And with Andry out of commission, I would need to find another field assistant to help me navigate the local politics. I felt the weight of the world on my shoulders. This was my PhD—I had made a commitment and sacrificed a great deal to come here. It felt like my entire project had fallen to pieces. I took a deep breath. Then another. If I can't do research, I thought, surely, I could travel. Do some reconnaissance. But I hadn't budgeted for a four-week excursion in Madagascar. What would that run me with hotels, food, and transportation?

Should I try to change my flight? Shawn had told me that it was not only difficult but expensive to change flights in Madagascar, because usually the flights to and from Paris were full. There I was: adrift in Madagascar with no data to show for my weeks in the country, nowhere to go, and nothing to do. Worst of all, I didn't have a research site to which I could return. My whole project had been a bust. I had nothing to show for all my hard work. Then, whoa! Enough! Things could have been far worse. What if, instead of Andry, I myself had been laid low with malaria? Rendered delirious? I decided then and there, even as we drove, that I would stop feeling sorry for myself. I would sort this out. I would find a way forward.

CHAPTER 14

Like the Chameleon

June 27, 2006

I called Fanja at La Maison du Pyla from the car as we neared Tana to see if she had a room available. Length of stay: indefinite.

"Oh no," she had said when I told her an abbreviated version of what had happened. "Yes, of course we can accommodate you, Keriann. Would you like the room Shawn had, with the en suite bathroom?" Let me think . . . *Yes*!

When we reached Tana, we dropped Andry off at his home. The driver pulled up out front and we all got out.

"Can you please unload Andry's luggage," I said, "and I will help him in."

Before we got out of the vehicle, I made Andry promise to update me as soon as possible. He thanked me quietly—he was still weak. Andry's uncle, a doctor, greeted us at the front door, and ushered Andry into the house, thanking me for my assistance. I didn't go in. I could see that Andry was in good hands.

It was two in the afternoon when I arrived at Fanja's—my sanctuary. I nearly collapsed in relief when she opened the door. Fanja made me some citronelle tea ("It will calm your nerves," she said) and we sat at the large round dining table in the main room while I told her the full story of what had happened. Her eyes widened at every detail, and as I finished the story she said, "Wow. You and Shawn both had such bad luck."

"That's true," I said, at first thinking she was referring to our shared journey into Kasijy. Then I remembered: Shawn's text. *Do not drive at night.* "Wait—did something happen to Shawn on his way out?"

Fanja raised her eyebrows. "You don't know?"

"I haven't yet connected with Shawn. When I tried to reach him, he was already on his way back to Canada."

"It was bad, Keriann." Fanja said, and she filled me in on Shawn's journey out of Kasijy. As he had said in his text, his hike out along the newly discovered (for us) trail had taken three days. At around four p.m., he had arrived in Kandreho, where the MICET truck was waiting. This trip had not only tested my limits but his too, and he was itching to get back to Tana. Sahoby, worried about his father, was equally anxious. And so, rather than spend the night in Kandreho, and begin driving next morning, Shawn, Sahoby, and the MICET driver left immediately. They intended to drive directly to Tana in one shot—no stops. This meant they would drive mostly in the dark.

Out on the lightless highway—like a rural country road in Canada—they approached a taxi brousse, driving slowly in front of them. The driver moved into the left lane to go around the slow-moving bush taxi, only to be confronted with a fast-moving oncoming vehicle. The MICET truck had nowhere to go. The oncoming car swerved and ended up in the ditch. It was a very close call. The MICET driver moved back into the right lane. But he didn't pull over to check on the driver of the other car. Shawn, now very shaken, begged the driver to stop. But the driver was afraid. He was afraid of what would at best be a confrontation with the driver of the other vehicle. He was afraid that the gendarmes would arrest him and, maybe most of all, he feared that he would lose his license and his job with MICET. And so, despite Shawn's frantic protests, he just kept driving.

A few kilometers down the road, a gendarme pulled them over. He had been alerted by the driver of the vehicle that had wound up in the ditch. Nobody was hurt, but the MICET driver was in big trouble for failing to stop at the scene. The three men spent the night in the gendarme office, despite their best efforts to smooth over the situation. In the end, MICET would sort everything out, but to Shawn and me, this was just another manifestation of what we later referred to as The Curse of Kasijy.

My Toronto travel doctor would have been appalled at Shawn's decision to drive a Madagascar highway at night, but I could empathize. We had both been through the ringer. Shawn's final misadventure, though terrible, could have been infinitely worse.

After Fanja and I had caught up, she showed me to my room—the one with the private bathroom. I took a long, hot shower. I was astonished by the amount of dirt that flowed off my skin. I scrubbed my body with soap and washed my hair twice with shampoo. And I conditioned. Oh, I conditioned all right. When, finally, I forced myself to exit the shower—I had to leave some hot water for the other guests—I felt human again.

I crawled into bed—I still had a few hours before dinner. I pulled out my laptop and began drafting emails: to Travis, to my parents, and to Shawn. Today was June 27. I was scheduled to fly home to Toronto on August 9. That meant that I had more than one month left in Madagascar—but a month to do what, exactly? I could feel anxiety creeping up on me. Before it could take over, I opened a new Excel spreadsheet on my laptop and set about working through my options.

First, I created rows for the hurdles that I was facing:

One month left in Madagascar
No viable field site to return to for my PhD research
Research permits are good only for Kasijy
Not much money left for traveling

Next, I created columns for possible solutions and next steps, corresponding to each of the hurdles:

Can I change my flight home? Visit Air France office to see what is possible.
Can I use the month to find a new field site and change my permits to conduct preliminary research elsewhere? Talk to Shawn and to Benjamin at MICET about this possibility.
Can I use the month to travel on the cheap? Consult my Lonely Planet guide.

Okay, I thought, *I have options here.* I felt better already. I put down my laptop and closed my eyes. I remembered some advice I had received from one of my graduate instructors at the U of T, when I told him I was headed for the field: "Remember, no proposal survives the field," he warned. "Something will inevitably not go as planned. You have to be able to adapt and revise." That put me in mind of another Malagasy proverb: "Like the chameleon—one eye on the future, one eye on the past."

Next morning, I tried and failed to reach Travis. But I connected with Shawn via Skype. I gave him an update, and he exclaimed, "What a mess this trip has been!"

We talked about what I could do, and he agreed that it would be worth checking with MICET. "There was another place in the northwest that I was thinking about working in when I first started looking for sites," he said. "Ankarafantsika National Park. It's a combination tourist and research site and you can access it easily—right off the highway."

Are you kidding me? A site right off the highway? That sounded good to me, especially at this point. Shawn doubted that I would be able to get a research permit but said the site was worth investigating, even if I visited as a tourist.

Next, I called my parents.

"Maybe you should just come home," my mom said. "Cut your losses?"

That sounded good too. I set out on a two-part investigation.

The woman at the Air France office told me it would be impossible to change my ticket. While I could change the date on the Paris ticket—there were seats available from Tana to Paris—the next leg of the journey, Paris to Toronto, was making things difficult. No seats available.

"Could I fly to Paris and then try for a standby flight?"

No. Air France would not send me to Paris unless I had a connecting flight to Toronto, as per my original ticket itinerary. I was stuck.

At MICET I talked with Benjamin. He agreed that Ankarafantsika would be a good place to visit. But to secure a research permit would take two or three months. Even if I could get a permit, he pointed out, I would be required to take a new student with me, which meant buying more gear and paying lab fees. I would also need new supplies, and to hire guides and a cook. I ran the math in my head. I would spend several hundred dollars . . . and for what? Two weeks of data at best? Like Shawn, Benjamin suggested that I might want to visit Ankarafantsika as a tourist. Go for a week, see the forest and the lemurs, and talk with the national park officials about coming back next year. He could arrange a driver for Tuesday.

Already, after just one smoggy day in Tana, I was itching to get back into nature. But if I left Tuesday and spent one week in Ankarafantsika, I would be back in Tana on July 11, which still left three full weeks in Madagascar—floating aimlessly. Sometimes I have trouble "rolling with the punches," as my father would say. I like plans, order, and a clear sense of what comes next. The idea of spending three full weeks without an itinerary made me crazy.

But if Ankarafantsika kept me occupied for the next week, maybe I could visit a few other national parks or reserves as a tourist. With Benjamin, I arranged for a vehicle to take me north on Tuesday, July 4.

Back at Fanja's I updated my parents. They had been talking with my travel agent, Mandy, who had booked my original ticket (this was 2006, so online tickets were not yet a thing). She suggested that I try again with the folks at Air France to see if they would let me change my ticket from Tana to Paris. Then, my parents could book the second part of the trip through another airline, maybe flying through London or Amsterdam. That option, however, would run me about $2,000.

My father said, "Maybe try again for a standby ticket?"

And so, the next day—Friday—I went back to the Air France office. The woman behind the counter recognized me and raised her eyebrows. I politely asked in my best French if there might be seats available for a flight home today. She shook her head no. And no, she couldn't send me without a ticket booked—standby would not cut it.

As I left the Air France office, I felt like crying. Not because I couldn't get home, though certainly I was beginning to feel homesick. It was because this was another failure. Turned out I had been right all along—I was an imposter, and here and now I was finally exposed. As I rode in the taxi back to Fanja's, I questioned all my decisions. Had I been too hasty with the evacuation? Maybe I should have stayed in Kasijy and just sent Andry out. That way, I would still have gotten some data and wouldn't be in this predicament. I wondered if that is what Shawn is thinking, too.

Back at the hotel, I managed to reach him on Skype.

"Go to Ankarafantsika," Shawn said. "And bring the satellite phone with you. If you find that you want to stay longer than a week, go for it—just phone MICET to let them know."

Then I could return to Tana. If I still felt like heading home, I could book a new flight.

"I can find you some funds to help pay for a flight change if you want," Shawn told me. "But, you know, it might be better to use those funds to check out a few other places in Madagascar, since you are there."

He mentioned Andasibe National Park, where I could see the Indri. "It is amazing," he said. "Really easy to see them, and you will get to hear them call." I flashed back to my very first anthropology class in Calgary, where Brian Keating had played the Indri's call for us. "It is just a short drive from Tana and really worth the trip," Shawn said. "There's also Kirindy National Park."

After I hung up, I pulled out my guidebook and lemur book and pored over the places I could visit. Slowly, my frustrations melted away. I started to get excited again. A trip around Madagascar could be a lot of fun. And if Ankarafantsika did end up being the place I would come back to, going to check it out was, in fact, work—I would accomplish something after all.

In my *Lemurs of Madagascar* book, Ankarafantsika gets special mention as one of the "Key Lemur-Watching Sites in Madagascar." Located forty-five kilometers northeast of Kasijy, Ankarafantsika National Park spans about 130,000 hectares. Like Kasijy, Ankarafantsika National Park is a dry deciduous forest. The soils are sandy. The most visited side is next to Ampijoroa village, directly off the highway. "The Ampijoroa campsite has been upgraded recently and now contains the locally managed Gite d'Ampijoroa with comfortable rooms and a restaurant."[1]

After where I had been, this sounded downright luxurious. A restaurant? But despite the amenities, I felt a pang of disappointment. Ankarafantsika was definitely not off the beaten track. Many a researcher and even tourists had visited that site. There had been something romantic about heading deep into the wilds of Madagascar to Kasijy, where only a few research teams had been. We had found some of the least-studied primate populations in the country. If only we could have stayed, who knows what I might have discovered. We could have made big contributions to primate research and conservation.

I shook it off and decided to read on:

> With a little bit of luck, one can see all eight lemur species in an
> afternoon, an evening, and a morning. Diurnal lemurs: the brown

lemur, the mongoose lemur, and Coquerel's sifaka. Nocturnal lemurs: the golden-brown mouse lemur, the gray mouse lemur, the
fat-tailed dwarf lemur, Milne-Edwards' sportive lemur, and the
Western woolly lemur.[2]

Okay, I reasoned, there are some entirely different lemurs than those I had seen
in Kasijy. I could add them to my lemur list. Now that I thought about it, I
could hardly wait.

The drive out to Ankarafantsika proved uneventful, unlike our previous
misadventures. But we were ten hours on the road, first taking Route Nationale
6 as we had done before, and then connecting to another major paved highway,
Route Nationale 4. We were just two in the old blue MICET van—me and
my driver, Pierre. I watched out the window as we wound through the rolling
Madagascar hills, passing through small towns and markets. We had left early—
around six a.m. It is so much easier to hit the road on time when you are alone.
We neared the park around four p.m., with dusk soon to settle in.

First, we passed through Andranofasika, the large community nearest the
park. This is where children from nearby villages go to school, Pierre said, as
he slowly navigated through the crowds. Several young men were out playing
basketball. Taxi brousses were parked at the side of the road, with people loading
them up with supplies. I spotted a few stores. And we were just two kilometers—
walking distance—from Ankarafantsika.

Pierre had no sooner announced this than we found ourselves surrounded
by forest. Yes, I thought, we must be close. I sat up a bit straighter in my seat.
Pierre slowed the van and I spotted a brown wooden sign. In large white block
letters, it said: *Parc National Ankarafantsika*. We had made it.

Pierre signaled left and we entered a gravel parking lot. Here I saw several
concrete buildings, the forest behind them. These must be the restaurants and
bungalows, I thought. We drove through a large archway—almost like one you
would see heading into a summer camp, painted with the camp's slogan. Atop
the archway, in large block letters, it read: *Royaume des oiseaux. Terre des lacs
sacrés. Source de vie.* (Kingdom of the birds. Land of the sacred lakes. Source of
life.) Something caught my eye—a flash of white and maroon atop the arch. It
couldn't be . . . could it? In fact, it was! A Coquerel's sifaka, right there in the
parking lot.

"Look, look, look!" Pierre pointed to some large mango trees behind the
archway.

There they were: seven sifakas, snacking on some mango fruits, bounding
from tree to tree, and scampering across the archway—and I hadn't even left the
vehicle. Lemurs in the parking lot. I couldn't help but laugh out loud.

Most field sites in Madagascar that you work in won't have electricity, so it is imperative that you carry batteries.

The Ampijoroa campground is divided into two sides: one for researchers and the other for tourists. The tourist side features small cabins, or gites, with electricity, single beds, and a shared bathroom (with flushing toilets, might I add). Here, too, you find several larger bungalows with multiple bunk beds, electricity, private bathrooms, and decks facing out onto the lake. Budget travelers (like me) can also set up a tent on one of the many tent platforms located near the restaurant building.

Also, nearby, as with any campground in Canada, are a series of washrooms and cold-water showers. I would spend a full week in Ankarafantsika, living out of my tent in the tourist campground, and eating at the local restaurant for breakfast, lunch, and dinner. The restaurant was overpriced by Malagasy standards (about US$5 per meal), and certainly not where the researchers would eat for months at a time, but I could see it becoming a nice occasional treat. At the national park office, I arranged for a tour guide to show me around for the week.

Andrema spoke good English and had a calm manner. "Tomorrow we will do a lemur tour," he said. "Then the next day I will show you the baobab and take you on a night walk." The five days following would include a boat ride, more lemur and bird-watching tours, and a trip to a large canyon. I told him I would also like to see the research camp and meet the director of research: "I might come back here as a researcher."

Andrema nodded and said we could do that too.

After dining alone in the restaurant—zebu brochette and French fries—I headed back to my tent and made a valiant effort to become engrossed in my book. But it wasn't happening. Although I was glad to be back in a forested area, and in my potential future field site, I was still fighting the nagging sense that I was a failure. In my journal from that first night in Ankarafantsika I find the words, I MISS TRAVIS, scrawled in large block letters at the bottom of the page.

Over the next few days in Ankarafantsika, I rode the Madagascar roller coaster of emotions. I was thrilled to be seeing lemurs and forest, frustrated that I couldn't collect any data, excited that this might be the place I would come back to, and annoyed that I had yet to see the research camp and meet the woman in charge of researchers. Every morning I would ask Andrema if today we could meet Madame Jacqueline, the head of research for Ankarafantsika National Park. Oh yes, he would say each time, but somehow the meeting never happened.

I couldn't complain too much because he filled my days with lemurs. I saw mongoose lemurs, Milne-Edwards's sportive lemurs, Coquerel's sifakas, and mouse lemurs. I saw lemur adults and lemur babies. I saw groups of lemurs and lemurs flying solo. And primates weren't all the park had to offer. On the baobab circuit I saw incredible birds like the sickle-billed vanga. Endemic to Madagascar, it is characterized by its three-inch-long, curved bill (shaped like a sickle), which it uses to pry bark off of trees, so it can get at its insect-prey.[3]

"When they call," Andrema said, "it sounds like a baby crying."

Kwaah Kwaah

Then there was the crested coua, also endemic to Madagascar. This terrestrial bird has bright blue skin around its eyes, so it looks like it is wearing 1980s-style eye shadow. Andrema said this bird is also called the timekeeper bird in Madagascar because it calls every hour on the hour. I spent the rest of that walk testing the validity of this assertion and decided that it might be close . . . but I wouldn't rely on it.

As we walked, Andrema pointed out more unique bird species, some found only in Madagascar, others only in this forest. During that week, I would see the Madagascar fish eagle, Madagascar harrier-hawk, the white-breasted mesite, the Madagascar grebe, Van Dam's vanga, and every bird-watcher's dream, the Schlegel's asity. Endemic to Madagascar, male Schlegel's asities are known for their large, light green and blue caruncles (fleshy outgrowths) that make it look as though they have a flower pattern around their eyes.[4] Bird-watching has never been my thing—I don't have the patience for it—but I found myself getting excited to see these brightly colored birds. Travis, an avid bird-watcher, would be in his glory here.

At the end of the baobab circuit, unsurprisingly, we came upon three baobab trees—the *Adansonia madagascariensis* species of baobab, said my guidebook. Baobabs are famous for their long life span (some have been around for thousands of years). They are huge (up to thirty meters tall and ten across), with bizarre crowns and swollen trunks that can store many gallons of water. People call them "upside-down trees" because their crowns sprawl out like roots. I would learn later that the three baobab trees at the end of the baobab circuit were of a species that is endemic to Madagascar, and the three trees that stood before me were the last remaining three of this subspecies, *Adansonia madagascariensis boenensis*.[5]

On the boat tour of Lac Ravelobe, Andrema described the rich cultural history of the lake. He explained that after the Merina tribe annexed the Sakalava people from the territory in the early 1800s, the proud royal family jumped into the lake and were eaten by crocodiles.[6] As a result, the local people consider the lake sacred. Each year, they pay tribute to their royal ancestors by sacrificing a zebu to the crocodiles, an action that grants them safety while fishing on the lake.[7] We spotted a large Nile crocodile sunning itself beside the lake. Andrema, who was not originally from the area, confided that because the lake and the crocodiles in it are considered sacred, the population density of the crocodiles is extremely high, and fishing next to the lake is very dangerous. Almost every year, he said, a crocodile eats a person or a few zebu.

On my last day in Ankarafantsika, Andrema took me to see the canyon. We made our way through the forest, along a wide trail made of loose grayish-white sand. The Grande Piste, as it was called, certainly was "grande." It was cut and maintained by the National Park, with the purpose of allowing guides to walk side by side with tourists as they looked for wildlife. As we neared the edge of the forest, the ground became firmer and the soil reddish in color, indicating its lateritic composition (high in iron).

"It's good to see the edge of the forest," I told Andrema as we walked. "If I come back here, I will be studying edge effects."

We transitioned from the forest to the grassy savannah. I was glad to see that the edge of the forest was only about five kilometers from the campsite, accessible along a large trail. My project would be infinitely easier to run here than it would have been in Kasijy. Next, we navigated through the savannah, making our way along a dirt road. The savannah here consisted of shorter grasses, red soils, and a few small scrubs. Unlike in Kasijy, we didn't have to worry about large boulders hidden in the tall grass. As we walked, I thought, *This is fantastic! The perfect place for my project.*

After a thirty-minute hike through the hot savannah, the layered, multicolored canyon came into view. It was beautiful.

"You would like a picture?" Andrema reached for my camera. A seasoned tour guide, he knew a photo op when he saw one. I handed over my camera and squatted down in front of the canyon. I was fighting mixed feelings about smiling. The canyon represented a gigantic erosion scar—evidence of human influence. I knew that wind and rain had created this canyon because of a lack of tree protection. At one time, this area was probably all forested.

Now, we looked out over a giant hole in the ground, a chasm that would forever prevent the forest on either side from regenerating and connecting. No trees grew in the canyon. Nothing could live there. Nothing could cross. How could we prevent this level of destruction in future? To figure that out, I would need to return. I knew now that Ankarafantsika was where I wanted to work.

I asked Andrema the same question I had asked every day since arriving: "Can you please take me past the research camp today?"

He must have understood that today I meant business. "Yes," he replied, finally.

We made our way back down the Grande Piste. But this time, instead of veering left toward the tourist camp, Andrema brought me right, down a secluded, unmarked pathway. We emerged in front of six or seven tent plat-forms—the same style as the ones in the tourist camp. Tents were set up on four of the platforms, with clotheslines strung up between the platform posts, shirts and pants drying in the sun. Below the tent platforms were two larger platforms that sat side by side and had concrete floors and thatched roofs. Each sheltered a large wooden table and chairs. On one of the platforms, someone had stacked pots and pans neatly on smaller tables. On the ground, I saw two green gas can-isters affixed with stovetops, and recognized this as the kitchen area.

Andrema led me deeper into the campsite. The place was empty.

"Where is everyone?" I asked.

"They must be working," Andrema replied.

He showed me the washroom area—three stalls with flushing toilets and three with cold-water showers. *This is fabulous*, I thought, *much more to my liking than the rustic Kasijy*. I could see myself here. I wanted to return for my PhD project. There was just one more thing: "Andrema, I need to speak with the head of research."

Madame Jacqueline was a petite, quiet woman with short dark hair. I shook hands with her and we sat in the main park office. She spoke very little English, so I did my best in French, explaining that I was "une chercheuse Canadienne," and that I wanted to come back in a year to study Coquerel's sifaka.

She told me about the required permissions and gave me a sheet of paper that broke down the camping and research fees for foreigners and Malagasy researchers. Finally, I had all the information I would need to return. I had seen the forest and the lemurs. I had found and explored the forest edge. And now, on my last day in Ankarafantsika, just in the nick of time, I had gathered the requisite information about permits and fees. And then, like kismet, through the window I saw the MICET truck pull in. We would spend one last night in Ankarafantsika and hit the road to Tana early next morning. My work here was done.

On the long drive back to Tana, I reflected that I had accomplished all I could. I was still mentally exhausted from my Kasijy ordeal. I had enjoyed my time in Ankarafantsika, but now I didn't feel much like traveling around by myself. Eating dinner alone each night and retreating to my tent was lonely. I didn't want to put in another three weeks. Sure, I could make the best of it, but really, I didn't want to. I decided that tomorrow, I would try one last time with Air France.

Next day, when she saw me coming, the woman behind the counter raised her eyebrows. I started to explain my situation, but she cut me off. She remem-bered, she said. I asked politely if she could please check to see if there were any available seats now. I braced myself for disappointing news. Only this time, after the woman keyed in my flight details, she looked at the screen, looked up at me, and paused. Then she smiled: "There is a seat."

My heart leapt. "A seat? For when?"

"Tonight."

I almost burst with excitement. A single obstacle remained. "How much will it cost to change my ticket and get on that flight?"

The woman again looked up from her computer screen, a gleam in her eye. With the slightest of smiles on her face, she said, "Aujourd'hui, c'est gratuit" (Today, it's free). Free! I did not know why and saw no need to inquire further.

I stumbled out of the Air France office clutching my new ticket. I could hardly believe it. I had found my new field site in Madagascar, ticked all the boxes there, and now I was heading home. The rest of the day was a blur. My flight would depart just before midnight, and I had a few loose ends to tie up. I visited the MICET office, expressed my delight with Ankarafantsika, and said I would fly out that evening. They said they could send a driver to take me to the airport. I promised to submit my final report from our work in Kasijy—required by Madagascar National Parks—via email from Canada.

Next, I stopped off at the Jumbo Score where I bought a sampling of Robert's chocolate, some vanilla beans, and a few souvenir T-shirts. My loot in hand, I made my way back to La Maison du Pyla to let Fanja know about my change of plans. When I arrived at the gates, the guardian informed me that I had a guest. He was waiting for me inside. Andry!

"Manahoana, Keriann." Andry stood to greet me as I walked through the door. I barely recognized him from the last time I had seen him. He was wearing a suit and tie and fairly sparkled.

"Andry!" I exclaimed, excited and relieved to see him looking so alive and well. "I'm so glad to see you. How are you feeling?"

Andry extended his hand to me. "I've come to thank you," he said with a serious tone.

"You don't need to—" I started.

"Keriann," he waved me off. "I want to thank you for saving my life."

CHAPTER 15

Planet Madagascar

In the months following my field research trip to Kasijy, my father kept bugging me to write about my experience. "You need to write this story." But I knew I wasn't ready. I was still questioning all my decisions. On that last day in Tana, Andry told me that his uncle, the doctor, had confirmed that he had malaria and that he was anemic. This was no surprise, as the majority of malarial infections are also associated with some degree of anemia.[1]

"I was reading about malarial anemia," my father said when I tried to brush him off. "It can cause death if the person is infected with falciparum malaria."

"But we didn't know whether Andry had falciparum," I said. "Things may not have been all that dire. Maybe Andry could have spent a week in Kandreho recovering, and we might then have been able to return to Kasijy."

"Yes, but only if you decided to overrule a medical doctor, a trained professional with local knowledge, who told you to take Andry back to Tana. You did the only right thing. You saved that young man's life. He thanked you for it himself. You're a hero and you can't seem to accept it."

I didn't feel like a hero. In the weeks after I got home, I went apartment hunting with my mother. In the midst of a Toronto heat wave, we pounded the pavement together until we found an apartment on Balsam Avenue in the Beaches—a first shared abode for Travis and me. It was blocks away from my parents' house—a big selling feature—and we would live there for a year before we flew to Madagascar, so I could start my full PhD project. On that second trip, Travis came with me.

A couple of weeks before we moved into the apartment, I found myself fighting headaches and nausea, and also having digestive issues. Something wasn't right. The doctor at the university clinic ran a few tests for tropical diseases, including malaria, just to be on the safe side. I received no call from the clinic, so figured it was a false alarm. Maybe I was just adjusting to being back in Canada.

Moving day came and, with Travis back in Belize, I found myself excitedly setting up house. I was unpacking clothes when a wave of nausea and dizziness hit me. I had to lie down. Something really wasn't right. I was reminded of the way that I felt on my first attempt at a nocturnal survey back in Kasijy. I just felt . . . off. I returned to my doctor and asked her to refer me to the Tropical Disease clinic at the Toronto General Hospital. If anyone could get to the bottom of this, it was Dr. Jay Keystone and his team. Dr. Keystone is an expert in tropical and infectious disease and a professor in the Department of Medicine at the University of Toronto. At his office, I saw the resident doctor. I described my symptoms, and he ordered more tests, taking twice the amount of blood they had taken at the university clinic. A few days later, I got an early-morning phone call from the medical lab. I had malaria.

"You need to get to the doctor immediately," the lab technician said, a hint of anxiety in his voice. I was probably the first malaria victim he had come across. "Tell them that you have vivax malaria."

I boarded the subway and headed to the university clinic that morning—a 45-minute transit ride. By the time I arrived at the clinic, I was feeling rather horrible. I had chills and all I wanted to do was lie down. I managed to stay standing long enough to check into the walk-in clinic. I told the woman at the desk: "I got a call this morning—I have vivax malaria." She gasped, unable to contain her surprise at my unusual predicament, and made a note in my file. After about an hour spent waiting, it was finally my turn. The nurse led me into the examination room where I waited again for the doctor to arrive.

The male doctor who saw me was tall and lanky, pale with dark brown hair. He walked in, holding my chart.

"What's this about malaria?" he asked, as he took his seat at his computer, a look of concern on his face.

I explained the situation. I had gotten a call from the lab. I have vivax malaria.

"Which lab?" he asked.

"Well, someone from the Tropical Disease Centre, I assume," I told him, weakly.

The doctor excused himself—he needed to check on something. When he returned he had a serious look on his face.

"Here's the deal," he said matter-of-factly. "Before we are able to prescribe malaria medication, the lab has to report the diagnosis into the Department of Health. We then have to wait for the Department of Health to call us and confirm the strain of malaria that you have."

I protested and shook my head: "But the lab tech told me—I have vivax malaria."

"I know," the doctor said sympathetically. "But I really can't prescribe anything for you. Hopefully they will call us soon. In the meantime, you can wait out on the nurse's bench."

I was ushered to the uncomfortable wooden nurse's bench. I waited. After about an hour the doctor who had seen me approached. I sat up straight in anticipation.

"My shift is over," he announced. "I am heading home now, but I filled in my colleague on the situation." *You've got to be joking.*

Finally, after another hour of waiting, the new doctor called me back and announced: "You have vivax malaria." *No kidding.* I collected a prescription for chloroquine, and then fetched the pills. I went to spend the night at my parents' house so that my mom could keep an eye on me. I had started my malaria treatment, but too late. The symptoms had taken hold hard. First, the chills. I had never been so cold. My teeth chattered uncontrollably. I did everything I could to get warm, including taking a hot bath. A few hours of that and then I was too hot. My head felt on fire. I had a nasty headache, ached all over my body, and felt like throwing up. My mother made up a pullout couch for me so that I could lie down. She took a cool washcloth to my burning forehead, but nothing could relieve this fever. I endured for a few more hours and finally the fever broke. What a relief. But my muscles and joints continued to ache and throb. I felt as if I had just run a marathon.

This must be how Andry felt, I thought, during our evacuation from Kasijy. Unbelievable. At least I was at home, in bed, with someone to take care of me. Plus, I had seen a doctor and was already on the necessary medication. Recovery was imminent, and I was just a short drive away from the nearest hospital if things took a nasty turn. Andry hadn't enjoyed any such luxuries. As I lay in bed, recovering from my own battle with malaria, I came around to my father's point of view. Maybe I had done the right thing after all.

Yet still I resisted writing about my experience. I didn't want to talk about what had happened, not even with my family and close friends. No way I wanted to commit the saga to as permanent a medium as the printed word. I couldn't shake the idea that I had failed. My preliminary research trip, which had cost over $5,000, had been a bust. I had returned to Toronto three weeks earlier than planned, and with no data collected. The only redeeming factor was that, having checked out Ankarafantsika, I had somewhere to go for my PhD project.

Ankarafantsika would become Shawn's permanent field site, where many of his graduate students would work. Starting in 2008, I lived and worked there myself, residing at the Ampijoroa research site for fourteen months. Much of the time, Travis was with me, deking back and forth to southern Africa to lead tours with an active-travel tour company called Backroads International—returning always with a bag of treats. We saw a lot during those fourteen months, and had more than a few memorable experiences, though none quite as dramatic as my journey to Kasijy.

Although I lived out of a tent, the workstation in Ankarafantsika was luxurious by fieldwork standards: electricity, flush toilets, showers, a restaurant on-site. Also, we were just an hour-long bus ride from the town of Mahajanga, where we would go on monthly supply runs. Then there were other field researchers working at the site—one team from Hanover, Germany, and another from Kyoto, Japan. We also met and became friends with an American couple from North Carolina, Megan and Dave, who spent a few months searching for the endangered mongoose lemur in northwest Madagascar.

Back in Canada, I buckled down to analyzing data and writing my dissertation. In 2009, Travis was accepted into the PhD program at the University of Toronto. The following year, he set up a remote satellite field camp near some forest fragments thirty kilometers from Ampijoroa. This site was easier to access than Kasijy but had no water source nearby. Travis coordinated a weekly water delivery, which involved a local community member, some jerry cans, and a zebu cart. He would return to that remote site for his full research project, visiting again in 2011 for seven months. I joined him that autumn for a couple of months, shortly after I defended my PhD thesis.

Our experiences in Madagascar changed both of us. The lemurs are disappearing, even in protected national parks like Ankarafantsika. While all the lemur species found in the park were present and accounted for near the main tourist site in Ampijoroa, we registered a sharp decline just outside that area, where Travis was working. The mongoose lemurs, for example, were nowhere to be seen, and the other large-bodied species, like Coquerel's sifaka, were few and far between. There just aren't enough extensive forest fragments for these species to survive. On top of that, the forest is burning. People are setting fires in the savannah to graze their cattle, and these fires sometimes encroach on the forest.

The challenges are more complex than we could have imagined. The people in Madagascar are desperately poor. Near Travis's field site, we visited three small communities, hamlets really. The people there live hand to mouth, so it is hard to fault them for burning forest when all they are doing is struggling to feed their families. During Travis's project, we saw people die needlessly from preventable ailments caused by poor sanitation and a lack of access to clean drinking water.

We decided that we couldn't stand by and watch as the lemurs became extinct and the people die. After we got back to Canada, Travis launched a project called *Lambas for Lemurs*. With our American friend Megan, he printed

Malagasy lambas with images of lemurs to raise funds for conservation education programs. The project was a success, and Travis decided to take it further. "I've got it," he said one day when I came in from a day of teaching. "You know how Madagascar is a world unto itself? I'm going to call my organization Planet Madagascar."

Now an official Canadian nonprofit, Planet Madagascar has run several community conservation projects in Madagascar, including a livelihoods survey, a conservation education program, and an ongoing fire management program. I sit on the board and help out with grant applications, but Travis is the one who has built this organization from the ground. In 2016, we received our first large grant, which enabled us to hire several new employees in Madagascar and one in Canada to help us to continue to grow. We have since started a forest restoration project and helped to set up a women's cooperative.

After completing my PhD, I spent a year applying for every tenure-track academic job that came up. Northern Illinois? Sure, I could live there. Kansas? Well, it wasn't on my radar, but okay. I even did a telephone interview with a university in Japan. To make ends meet, I went on what I call the "Southern Ontario Sessional Circuit." I taught undergraduate courses at the University of Waterloo, Trent University, and the University of Toronto—sometimes all at once. During one stretch, I found myself teaching five days a week, and commuting some distance by bus for four of them. Waterloo, Peterborough, Mississauga. No surprise, surely, that I started casting around for something different.

I ended up enrolling in a publishing program at Ryerson University. While I shunted between universities, I took a few courses, won an award, and made some connections in the industry. I did a couple of internships and finally got a foothold with a scholarly press. From there, I moved to a leading educational book publisher, where I worked as an editor.

Of course, I remain deeply committed to Planet Madagascar, which aims to conserve the island-nation's amazing and fragile biodiversity, while also helping and empowering the Malagasy people. We have developed a fire management program, and plan to tackle issues like women's health and sustainable agriculture. This involvement is what drove me finally to write this book. I kept thinking about the lemurs and the Malagasy people, and in the summer of 2016, I found myself flipping through the journal I had kept during my first brief sojourn in Madagascar. I thought of Sahoby, so reserved, and Fidèle whistling while he walked. I thought of Andry, who had dressed up in a suit to pay me a final visit. Something came over me, I don't know what. All I know is that I sat down at the keyboard and these words started pouring out of me.

Full Disclosure

This book tells a true story as accurately as possible. My trip in and out of Kasijy in 2006 happened as described. While writing, I revisited my journal and emails. There were times, of course, that I was unsure of the sequence of events or precise words, and I have changed a few names. I have done my best to stick to the facts, and have written as honestly as I could, telling the story as I remember it. My PhD supervisor did provide his students with a set of notes and travel suggestions, which he called the "TREE Team Guide to Madagascar." For the sake of readability, I have "improved" that guide.

THE END

Notes

Prologue

1. David Werner, Carol Thuman, and Jane Maxwell, *Where There Is No Doctor: A Village Health Care Handbook* (London: Macmillan Education, 1993).

Introduction

1. Christoph Schwitzer et al., eds., *Lemurs of Madagascar: A Strategy for Their Conservation 2013–2016* (Bristol, UK: IUCN SSC Primate Specialist Group, Bristol Conservation and Science Foundation, and Conservation International, 2013).

Chapter 1

1. Christoph Schwitzer et al., eds., *Lemurs of Madagascar: A Strategy for Their Conservation 2013–2016* (Bristol, UK: IUCN SSC Primate Specialist Group, Bristol Conservation and Science Foundation, and Conservation International, 2013).
2. Ibid.
3. Ibid.
4. Kew Gardens, "Country Focus—Status of Knowledge of Madagascan Plants," *State of the World's Plants, 2017*, https://stateoftheworldsplants.org/2017/report/SOTWP_2017_6_country_focus_status_of_knowledge_of_madagascan_plants.pdf (accessed March 31, 2019).
5. Ibid.
6. "Madagascar and Indian Ocean Islands—Species," *Biodiversity Hotspots*, Critical Ecosystem Partnership Fund, https://www.cepf.net/our-work/biodiversity-hotspots/madagascar-and-indian-ocean-islands/species (accessed March 31, 2019).

7. Ibid.

8. Schwitzer et al., *Lemurs of Madagascar*.

9. Russell A. Mittermeier et al., eds., *Lemurs of Madagascar*, 3rd edition (Arlington, VA: Conservation International, 2010).

10. Ibid., 70.

11. Ibid.

12. Ghislain Vieilledent et al., "Combining Global Tree Cover Loss Data with Historical National Forest Cover Maps to Look at Six Decades of Deforestation and Forest Fragmentation in Madagascar," *Biological Conservation* 222 (2018): 189–97.

13. David A. Burney et al., "A Chronology for Late Prehistoric Madagascar," *Journal of Human Evolution* 47, nos. 1–2 (2004): 25–63.

14. James Hansford et al., "Early Holocene Human Presence in Madagascar Evidenced by Exploitation of Avian Megafauna," *Science Advances* 4, no. 9 (2018): eaat6925.

15. The World Bank, "Madagascar: Overview," The World Bank in Madagascar, 2017, https://www.worldbank.org/en/country/madagascar/overview (accessed April 1, 2019).

16. Ibid.

17. "Water, Sanitation, and Hygiene," UNICEF Madagascar, https://www.unicef.org/madagascar/eng/wes_15164.html (accessed April 1, 2019).

18. Alison Jolly and Richard Jolly, "Malagasy Economics and Conservation: A Tragedy without Villains," in *Madagascar*, ed. Alison Jolly et al. (Oxford: Pergamon Press, 1984), 211–18.

19. "All about Jane," The Jane Goodall Institute of Canada, https://janegoodall.ca/who-we-are/all-about-jane/ (accessed April 1, 2019).

20. Ibid.

21. Jacob Dunn et al., "Evolutionary Trade-Off between Vocal Tract and Testes Dimensions in Howler Monkeys," *Current Biology* 21, no. 2 (2015): 2839–44.

22. "Belize Country Profile," BBC News, http://news.bbc.co.uk/2/hi/americas/country_profiles/1211472.stm (accessed April 1, 2019).

23. "Madagascar Country Profile," BBC News, https://www.bbc.com/news/world-africa-13864364 (accessed April 1, 2019).

24. "What Languages Are Spoken in Madagascar?" World Atlas, https://www.worldatlas.com/articles/what-languages-are-spoken-in-madagascar.html (accessed April 1, 2019).

25. Pauline R. Clance and Suzanne A. Imes, "The Impostor Phenomenon in High Achieving Women: Dynamics and Therapeutic Intervention," *Psychotherapy Theory, Research and Practice* 15, no. 3 (1978): 241–47.

26. John Steinbeck, *Working Days: The Journals of "The Grapes of Wrath"* (New York: Penguin Books, 1990), 56.

Chapter 2

1. Deborah Curtis et al., "Surveys on *Propithecus verreauxi deckeni*, a Melanistic Variant, and *P. v. coronatus* in North-West Madagascar," *Oryx* 32, no. 2 (1998): 157–64;

P. M. Randrianarisoa et al., "Inventaire des lémuriens dans la Réserve Spéciale de Kasijy," *Lemur News* 6 (2001): 7–8.

2. N. Andriaholinirina et al., *"Propithecus coronatus," The IUCN Red List of Threatened Species,* 2014, https://www.iucnredlist.org/species/18356/16115921(accessed April 6, 2019).

3. "International Travel and Health: Malaria," World Health Organization, https://www.who.int/ith/diseases/malaria/en/ (accessed April 2, 2019).

4. Shawn M. Lehman, Andry Rajaonson, and Sabine Day, "Lemur Responses to Edge Effects in the Vohibola III Classified Forest, Madagascar," *American Journal of Primatology* 68, no. 3 (2006): 293–99.

5. Peter M. Kappeler, "Lemur Origins: Rafting by Groups of Hibernators?" *Folia Primatologica* 71, no. 6 (2000): 422–25.

6. Jörg U. Ganzhorn and Jutta Schmid, "Different Population Dynamics of *Microcebus murinus* in Primary and Secondary Deciduous Dry Forests of Madagascar," *International Journal of Primatology* 19, no. 5 (1998): 785–96.

7. Shawn M. Lehman, Andry Rajaonson, and Sabine Day, "Edge Effects on the Density of *Cheirogaleus major*," *International Journal of Primatology* 27, no. 6 (2006): 1569–88.

8. Lehman et al., "Lemur Responses to Edge Effects."

9. Ibid.

10. Russell A. Mittermeier et al., eds., *Lemurs of Madagascar*, 3rd edition (Arlington, VA: Conservation International, 2010).

Chapter 3

1. Russell A. Mittermeier et al., eds., *Lemurs of Madagascar*, 3rd edition (Arlington, VA: Conservation International, 2010).

2. Ibid.

3. Ibid.

4. Ibid.

5. Ibid.

6. Ibid.

7. Laurie R. Godfrey, William L. Jungers, and David A. Burney, "Subfossil Lemurs of Madagascar," *Cenozoic Mammals of Africa* (2010): 351–67.

8. Ibid.

9. Laurie R. Godfrey, "Subfossil Lemurs," in *The International Encyclopedia of Primatology*, ed. Agustín Fuentes, 3 vols. (Malden, MA: Wiley Blackwell, 2016), 1348–51.

10. Laurie R. Godfrey, William L. Jungers, and Gary T. Schwartz, "Ecology and Extinction of Madagascar's Subfossil Lemurs," in *Lemurs: Ecology and Adaptation*, ed. Lisa Gould and M. L. Sauther (New York: Springer, 2006), 41–64.

11. Ibid.

12. Ibid.

13. Ventura Perez et al., "Evidence of Early Butchery of Giant Lemurs in Madagascar," *Journal of Human Evolution* 49, no. 6 (2005): 722–42.

14. Godfrey et al., "Ecology and Extinction of Madagascar's Subfossil Lemurs."

15. Laurie R. Godfrey et al., "A New Interpretation of Madagascar's Megafaunal Decline: The "Subsistence Shift Hypothesis," *Journal of Human Evolution* 130 (2019): 126–40.

16. Ibid.

17. Ibid.

18. Christoph Schwitzer et al., eds., *Lemurs of Madagascar: A Strategy for Their Conservation 2013–2016* (Bristol, UK: IUCN SSC Primate Specialist Group, Bristol Conservation and Science Foundation, and Conservation International, 2013).

19. Claudia Dreifus, "A Conversation With: A Lemur Rescue Mission in Madagascar," *New York Times*, 2014, https://www.nytimes.com/2014/08/19/science/a-rescue-mission-in-madagascar.html (accessed April 2, 2019).

20. N. Andriaholinirina et al., "*Hapalemur aureus*," *The IUCN Red List of Threatened Species*, 2014, https://www.iucnredlist.org/species/9672/16119513 (accessed April 7, 2019).

21. Kenneth Glander et al., "Consumption of Cyanogenic Bamboo by a Newly Discovered Species of Bamboo Lemur," *American Journal of Primatology* 19, no. 2 (1989): 119–24.

22. Timothy Eppley et al., "High Energy or Protein Concentrations in Food as Possible Offsets for Cyanide Consumption by Specialized Bamboo Lemurs in Madagascar," *International Journal of Primatology* 38 (2017): 881–99.

23. Andriaholinirina et al., "*Hapalemur aureus*."

24. Ibid.

25. Ibid.

26. Linda Marie Fedigan, "Science and the Successful Female: Why There Are So Many Women Primatologists," *American Anthropologist* 96, no. 3 (1994): 529–40.

27. Richard Wrangham, "Evolution of Coalitionary Killing," *Yearbook of Physical Anthropology* 42 (1999): 1–30.

28. Ibid.

29. Carel P. van Schaik and Peter M. Kappeler, "The Social Systems of Gregarious Lemurs: Lack of Convergence with Anthropoids Due to Evolutionary Disequilibrium?" *Ethology* 102, no. 7 (1996): 915–41.

30. Michelle Sauther, Robert W. Sussman, and Lisa Gould, "The Socioecology of the Ringtailed Lemur: Thirty-Five Years of Research," *Evolutionary Anthropology: Issues, News, and Reviews* 8, no. 4 (1999): 120–32.

31. Fedigan, "Science and the Successful Female."

Chapter 4

1. Dune du Pilat, https://www.dunedupilat.com/en/ (accessed April 2, 2019).

2. Pierre Manganirina Randrianarisoa, "Inventaire des lémuriens dans la Réserve Spéciale de Kasijy," *Lemur News* 6 (2001): 7.

3. Merriam-Webster, s.v. "sympatric," https://www.merriam-webster.com/dictionary/sympatric (accessed April 5, 2019).

4. Russell A. Mittermeier et al., eds., *Lemurs of Madagascar*, 3rd edition (Arlington, VA: Conservation International, 2010).

5. Ibid.

6. Ibid.

7. Ibid.

8. Ibid.

9. N. Andriaholinirina et al., "*Hapalemur griseus*," *The IUCN Red List of Threatened Species* 2014, https://www.iucnredlist.org/species/9673/16119642 (accessed April 9, 2019).

10. Ibid.

11. Ibid.

12. UN Department of Economic and Social Affairs, "The World's Cities in 2016, Statistical Papers," in *United Nations (Ser. A), Population and Vital Statistics Report* (New York: United Nations, 2016), https://doi.org/10.18356/8519891f-en (accessed on April 2, 2019).

13. *New World Encyclopedia*, s.v. "Antananarivo," http://www.newworldencyclopedia.org/p/index.php?title=Antananarivo&oldid=995044 (accessed April 2, 2019).

14. Travel Madagascar, http://www.travelmadagascar.org/CITIES/Antananarivo-what-to-see.html (accessed April 12, 2019).

15. The World Bank, "Madagascar: Overview," The World Bank in Madagascar, 2017, https://www.worldbank.org/en/country/madagascar/overview (accessed April 2, 2019).

16. "Water, Sanitation, and Hygiene," UNICEF Madagascar, https://www.unicef.org/madagascar/eng/wes_15164.html (accessed April 2, 2019).

17. Hilary Bradt, *Madagascar: The Bradt Travel Guide* (Chesham, UK: Bradt Travel Guides, 2007).

18. "XE The World's Trusted Currency Converter," https://www.xe.com (accessed April 2, 2019).

19. "GDP per capita (US$)," The World Bank, https://data.worldbank.org/indicator/NY.GDP.PCAP.CD (accessed April 2, 2019).

20. Chocolaterie Robert, https://www.chocolaterierobert.com (accessed April 2, 2019).

21. MadaCamp, "Chocolaterie Robert," https://www.madacamp.com/Chocolaterie_Robert (accessed April 2, 2019).

22. Chocolaterie Robert.

23. Ibid.

Chapter 5

1. Hilary Bradt, *Madagascar: The Bradt Travel Guide* (Chesham, UK: Bradt Travel Guides, 2007).

2. David A. Burney et al., "A Chronology for Late Prehistoric Madagascar," *Journal of Human Evolution* 47, nos. 1–2 (2004): 25–63.

3. James Hansford et al., "Early Holocene Human Presence in Madagascar Evidenced by Exploitation of Avian Megafauna," *Science Advances* 4, no. 9 (2018): eaat6925.

4. Denis Pierron et al., "Strong Selection during the Last Millennium for African Ancestry in the Admixed Population of Madagascar," *Nature Communications* 9, no. 1 (2018): 932.

5. Denis Pierron et al., "Genomic Landscape of Human Diversity across Madagascar," *Proceedings of the National Academy of Sciences* 114, no. 32 (2017): E6498–E6506.

6. Laurie R. Godfrey et al., "A New Interpretation of Madagascar's Megafaunal Decline: The 'Subsistence Shift Hypothesis,'" *Journal of Human Evolution* 130 (2019): 126–40.

7. Ibid.

8. Robert E. Dewar and Henry T. Wright, "The Culture History of Madagascar," *Journal of World Prehistory* 7, no. 4 (1993): 417–66.

9. Ibid.

10. Alexander Ives Bortolot, "Kingdoms of Madagascar: Maroserana and Merina," in *Heilbrunn Timeline of Art History* (New York: Metropolitan Museum of Art, 2000), http://www.metmuseum.org/toah/hd/madg_1/hd_madg_1.htm (accessed April 12, 2019).

11. H. T. Wright and J. A. Rakotoarisoa, "The Rise of Malagasy Societies: New Developments in the Archaeology of Madagascar, in *The Natural History of Madagascar*, ed. Steven M. Goodman and Jonathan P. Benstead (Chicago: University of Chicago Press), 112–19.

12. Bortolot, "Kingdoms of Madagascar."

13. Ibid.

14. John Middleton, *World Monarchies and Dynasties* (Abingdon: Routledge, 2004).

15. Dewar and Wright, "The Culture History of Madagascar."

16. Ibid.

17. Bortolot, "Kingdoms of Madagascar."

18. Steven L. Danver, *Native Peoples of the World: An Encyclopedia of Groups, Cultures and Contemporary Issues* (New York: Routledge, 2015), 61.

19. Dewar and Wright, "The Culture History of Madagascar."

20. "Rice around the World: Madagascar," Food and Agriculture Organization of the United Nations, International Year of Rice 2004, http://www.fao.org/rice2004/en/p9.htm (accessed April 6, 2019).

21. Matthew E. Hurles et al., "The Dual Origin of the Malagasy in Island Southeast Asia and East Africa: Evidence from Maternal and Paternal Lineages," *American Journal of Human Genetics* 76, no. 5 (2005): 894–901.

22. Dietrich Werner, ed., *Biological Resources and Migration* (Berlin: Springer, 2013).

23. B. Minten et al., "Rice Markets in Madagascar in Disarray: Policy Options for Increased Efficiency and Price Stabilization," working paper, World Bank, 2006.

24. Erika Styger et al., "Influence of Slash-and-Burn Farming Practices on Fallow Succession and Land Degradation in the Rainforest Region of Madagascar," *Agriculture, Ecosystems, and Environment* 119 (2007): 257–69.

25. Ibid.

26. "Population Density (people per sq. km of land area), 2019," The World Bank, https://data.worldbank.org/indicator/EN.POP.DNST?locations=CH (accessed April 2, 2019).

27. Styger et al., "Influence of Slash-and-Burn Farming Practices."

28. Christoph Schwitzer et al., eds., *Lemurs of Madagascar: A Strategy for Their Conservation 2013–2016* (Bristol, UK: IUCN SSC Primate Specialist Group, Bristol Conservation and Science Foundation, and Conservation International, 2013).

29. "Madagascar: Improving Farmers' Incomes," The World Bank, April 27, 2018, https://www.worldbank.org/en/about/partners/brief/madagascar-improving-farmers-incomes (accessed April 12, 2019).

Chapter 6

1. Hilary Bradt, *Madagascar: The Bradt Travel Guide* (Chesham, UK: Bradt Travel Guides, 2007).

2. Ibid.

3. "Natural Wild Silk: Making the Most of Madagascar's Biodiversity," Food and Agriculture Organization of the United Nations, 2010, http://www.fao.org/3/am036e/am036e09.pdf (accessed April 2, 2019).

4. Ibid.

5. Walter E. Little and Paula A. McAnany, eds., *Textile Economies: Power and Value from the Local to the Transnational*, Society for Economic Anthropology Monograph Series (Lanham, MD: AltaMira Press, 2011).

6. Maki and Mpho, "The Future of African Luxury Feat: 'African' Silk from Madagascar," July 11, 2014, http://www.makiandmpho.com/blog/2014/7/11/the-future-of-african-luxury-african-silk-from-madagascar (accessed April 3, 2019).

7. Ibid.

8. "Natural Wild Silk."

9. Brian Barth, "A Silk Road to Community Prosperity," *Earth Island Journal* (Summer 2015), http://www.earthisland.org/journal/index.php/magazine/entry/a_silk_road_to_community_prosperity/ (accessed April 2, 2019).

10. SEPALI, https://www.sepalim.org (accessed April 3, 2019); Food and Agriculture Organization of the United Nations, "Natural Wild Silk."

11. Laingoniaina Herifito et al., "A Preliminary Assessment of Sifaka (*Propithecus*) Distribution, Chromatic Variation and Conservation in Western Central Madagascar," *Primate Conservation* 28 (2014): 43–53.

12. P. M. Randrianarisoa et al., "Inventaire des lémuriens dans la Réserve Spéciale de Kasijy," *Lemur News* 6 (2001): 7–8.

13. Grady J. Harper et al., "Fifty Years of Deforestation and Forest Fragmentation in Madagascar," *Environmental Conservation* 34, no. 4 (2007): 325–33.

14. Norman Myers et al., "Biodiversity Hotspots for Conservation Priorities," *Nature* 403 (2000): 843–53.

15. Ibid.

16. "Biodiversity Hotspots: Targeted Investment in Nature's Most Important Places," Conservation International, https://www.conservation.org/How/Pages/Hotspots.aspx (accessed April 3, 2019).

17. Ibid.

18. Christoph Schwitzer et al., eds., *Lemurs of Madagascar: A Strategy for Their Conservation 2013–2016* (Bristol, UK: IUCN SSC Primate Specialist Group, Bristol Conservation and Science Foundation, and Conservation International, 2013).

19. The World Bank, "Madagascar: Overview," The World Bank in Madagascar, 2017, https://www.worldbank.org/en/country/madagascar/overview (accessed April 2, 2019).

20. Jenny Fuhr, *Experiencing Rhythm: Contemporary Malagasy Music and Identity* (Newcastle upon Tyne: Cambridge Scholars Publishing, 2013).

21. Ibid.

Chapter 7

1. "The Hoodoos of Drumheller Valley," Atlas Obscura, https://www.atlasobscura.com/places/hoodoos (accessed April 9, 2019).

2. Peter Tyson, *The Eighth Continent: Life, Death and Discovery in a Lost World* (New York: William Morrow, 2000), 224.

3. Ibid., 374.

4. Ibid., 223.

5. Ibid., 374.

6. Ibid.

7. Urs Bloesch, "Fire as a Tool in the Management of a Savanna/Dry Forest Reserve in Madagascar," *Applied Vegetation Science* 2, no. 1 (1999): 117–24.

8. Jørgen Klein, Betrand Réau, and Mary Edwards, "Zebu Landscapes: Conservation and Cattle in Madagascar," in *Greening the Great Red Island: Madagascar in Nature and Culture*, ed. J. C. Kaufmann (Pretoria: Africa Institute of South Africa [AISA], 2018), 157–78.

9. Thomas A. Green and Joseph R. Svinth, eds., "Moraingy," in *Martial Arts of the World: An Encyclopedia of History and Innovation* (Santa Barbara, CA: ABC-CLIO, 2010), 14–18.

10. Ibid.

Chapter 8

1. "Africa: Madagascar," The World Factbook, https://www.cia.gov/library/publications/the-world-factbook/geos/print_ma.html (accessed April 5, 2019).

2. Ibid.

3. Pradiptajati Kusuma, "In Search of Asian Malagasy Ancestors in Indonesia," PhD thesis, Université de Toulouse, 2017.

4. Peter Tyson, *The Eighth Continent: Life, Death and Discovery in a Lost World* (New York: William Morrow, 2000).

5. Ibid., 241.

6. Ibid.

7. Ibid., 281.

8. Ibid., 243.

9. Small Arms Survey, "Ethos of Exploitation: Insecurity and Predation in Madagascar," in *Small Arms Survey 2011: States of Security* (Cambridge: Cambridge University Press, 2011), 172.

10. Michael Lambek, *The Weight of the Past: Living with History in Mahajanga, Madagascar* (New York: Palgrave Macmillan, 2002).

11. Tyson, *The Eighth Continent*, 120.

12. Ibid.

13. T. M. Lejeune, Patrick A. Willems, and Norman C. Heglund, "Mechanics and Energetics of Human Locomotion on Sand," *Journal of Experimental Biology* 201 (1998): 2071–80.

Chapter 9

1. Beth Moore, "Remember Jellies? The Clear Plastic Shoes Are Back, Updated with Flair," *Baltimore Sun*, June 29, 2003, https://www.baltimoresun.com/news/bs-xpm -2003-06-29-0306290438-story.html (accessed April 5, 2019).

2. Alison Behie, Mary S. M. Pavelka, and Colin A. Chapman, "Sources of Variation in Fecal Cortisol Levels in Howler Monkeys in Belize," *American Journal of Primatology* 72 (2010): 600–606.

3. R. P. D. Walsh, "Climate," in *The Tropical Rain Forest*, ed. P. W. Richards (Cambridge: Cambridge University Press, 1996), 159–236.

4. Jorg U. Ganzhorn et al., "The Biodiversity of Madagascar: One of the World's Hottest Hotspots on Its Way Out," *Orxy* 35, no. 4 (2001): 1–3.

5. D. J. Overdorff, "Similarities, Differences, and Seasonal Patterns in the Diets of *Eulemur rubriventer* and *Eulemur fulvus rufus* in the Ranomafana National Park, Madagascar," *International Journal of Primatology* 14 (1993): 721–53.

6. Patricia C. Wright, "Considering Climate Change Effects in Lemur Ecology and Conservation," in *Lemurs* (Boston: Springer, 2006), 385–401.

7. Peter Tyson, *The Eighth Continent: Life, Death and Discovery in a Lost World* (New York: William Morrow, 2000).

8. Ibid., 107.

9. The Mad Dog Initiative, http://www.maddoginitiative.com (accessed April 5, 2019).

10. Hilary Bradt, *Madagascar: The Bradt Travel Guide* (Chesham, UK: Bradt Travel Guides, 2007).

11. Ibid.

12. Carl Safina, "The Legend of Babakoto," *National Geographic*, October 22, 2016, https://blog.nationalgeographic.org/2016/10/22/the-legend-of-babakoto/ (accessed April 12, 2019).

Chapter 10

1. Russell A. Mittermeier et al., eds., *Lemurs of Madagascar,* 3rd edition (Arlington, VA: Conservation International, 2010).

2. "Mouse Lemurs," Photo Ark, *National Geographic*, https://www.nationalgeo graphic.com/animals/mammals/group/mouse-lemurs/ (accessed April 10, 2019).

3. Shawn M. Lehman, Andry Rajaonson, and Sabine Day, "Edge Effects and Their Influence on Lemur Density and Distribution in Southeast Madagascar," *American Journal of Physical Anthropology* 129, no. 2 (2006): 232–41.

4. Mittermeier et al., *Lemurs of Madagascar.*

5. Ibid.

6. Steven M. Goodman, Olivier Langrand, and Christopher J. Raxworthy, "The Food Habits of the Madagascar Long-Eared Owl *Asio madagascariensis* in Southeastern Madagascar," *Bonn Zoological Bulletin* 42, no. 1 (1991): 21–26.

7. Ute Radespiel et al., "Sex-Specific Usage Patterns of Sleeping Sites in Grey Mouse Lemurs (*Microcebus murinus*) in Northwestern Madagascar," *American Journal of Primatology* 46 (1998): 77–84.

8. Hideyuki Doi and Teruhiko Takahara, "Global Patterns of Conservation Research Importance in Different Countries of the World," *PeerJ* 4 (July 2016): e2173, https://doi.org/10.7717/peerj.2173 (accessed April 4, 2019).

9. "Nile Crocodile," Photo Ark, *National Geographic*, https://www.nationalgeo graphic.com/animals/reptiles/n/nile-crocodile/ (accessed April 4, 2019).

10. Ibid.

11. Alina Bradford, "Crocodiles: Facts and Pictures," *LiveScience*, June 25, 2014, https://www.livescience.com/28306-crocodiles.html (accessed April 5, 2019).

12. "CITES Bans Malagasy Crocodile Product Trade," Worldwide Fund for Nature, (WWF), May 19, 2010, http://wwf.panda.org/?193342/CITES-bans-Malagasy-crocodile -product-trade (accessed April 5, 2019).

13. Simon Pooley, "A Cultural Herpetology of Nile Crocodiles in Africa," *Conservation and Society* 14, no. 4 (2016): 391–405.

14. "CITES Bans Malagasy Crocodile Product Trade."

15. Christoph Schwitzer et al., eds., *Lemurs of Madagascar: A Strategy for Their Conservation 2013–2016* (Bristol, UK: IUCN SSC Primate Specialist Group, Bristol Conservation and Science Foundation, and Conservation International, 2013).

16. Tsilavo Raharimahefa and Timothy Kusky, "Environmental Monitoring of Bombetoka Bay and the Betsiboka Estuary, Madagascar, Using Multi-temporal Satellite Data," *Journal of Earth Science* 21 (2010): 210–26.

17. Rattan Lal, "Soil Erosion and Land Degradation: The Global Risks," in *Advances in Soil Science, Volume 11: Soil Degradation*, ed. R. Lal and B. A. Stewart (New York: Springer, 1990), 129–72.

Chapter 11

1. Nick Garbutt, *Mammals of Madagascar* (New Haven, CT: Yale University Press, 2007).

2. Ibid., 247.

3. Russell A. Mittermeier et al., eds., *Lemurs of Madagascar*, 3rd edition (Arlington, VA: Conservation International, 2010).

4. Ibid.

5. N. Andriaholinirina, "*Eulemur rufus*," *The IUCN Red List of Threatened Species* https://www.iucnredlist.org/species/8209/16118167 (accessed April 5, 2019).

6. Ian Tattersall, "Cathemeral Activity in Primates: A Definition," *Folia Primatologica* 49 (1987): 200–202.

7. Giuseppe Donati, "Ecological and Anthropogenic Correlates of Activity Patterns in *Eulemur*," *International Journal of Primatology* 37, no. 1 (2016): 29–46.

8. Deborah J. Overdorff et al., "Life History of *Eulemur fulvus rufus* from 1988–1998 in Southeastern Madagascar," *American Journal of Physical Anthropology: The Official Publication of the American Association of Physical Anthropologists* 108, no. 3 (1999): 295–310; Mittermeier et al., *Lemurs of Madagascar*.

9. Russell Mittermeier et al., "Primates in Peril: The World's 25 Most Endangered Primates 2008–2010," *Primate Conservation* 24, no. 1 (2009): 1–58.

10. Ibid.

11. Andy Purvis et al., "Predicting Extinction Risk in Declining Species," *Proceedings of the Royal Society of London. Series B: Biological Sciences* 267, no. 1456 (2000): 1947–52.

12. Richard K. B. Jenkins et al., "Analysis of Patterns of Bushmeat Consumption Reveals Extensive Exploitation of Protected Species in Eastern Madagascar," *PloS One* 6, no. 12 (2011): e27570.

13. Christoph Schwitzer et al., eds., *Lemurs of Madagascar: A Strategy for Their Conservation 2013–2016* (Bristol, UK: IUCN SSC Primate Specialist Group, Bristol Conservation and Science Foundation, and Conservation International, 2013).

14. Christopher D. Golden, "Bushmeat Hunting and Use in the Makira Forest, North-Eastern Madagascar: A Conservation and Livelihoods Issue," *Oryx* 43, no. 3 (2009): 386–92.

15. Schwitzer et al., *Lemurs of Madagascar*.

16. Emily Graslie, "Periods + Fieldwork," *The BrainScoop*, February 10, 2016, https://www.youtube.com/watch?v=jjFZ1nzijrI (accessed April 5, 2019).

17. Rachel Becker, "Fighting the Menstruation Taboo in the Field," *Nature*, February 12, 2016, https://www.nature.com/news/fighting-the-menstruation-taboo-in-the-field-1.19372 (accessed April 5, 2019).

18. Graslie, "Periods + Fieldwork."

19. Michael Lambek, *The Weight of the Past: Living with History in Mahajanga, Madagascar* (New York: Palgrave Macmillan, 2002).

20. Mittermeier et al., *Lemurs of Madagascar*; D. J. Overdorff and S. Johnson, "*Eulemur*, True Lemurs," in *The Natural History of Madagascar*, ed. Steven M. Goodman and Johnathan P. Benstead (Chicago: University of Chicago Press, 2003), 1320–24.

21. Overdorff and Johnson, "*Eulemur*, True Lemurs."

22. Ibid.

23. Ibid.

24. Roger A. Powell and Michael S. Mitchell, "What Is a Home Range?" *Journal of Mammalogy* 93, no. 4 (2012): 948–58.

25. Chris Carbone et al., "How Far Do Animals Go? Determinants of Day Range in Mammals," *American Naturalist* 165, no. 2 (2004): 290–97.

26. Colin Chapman, "Patch Use and Patch Depletion by the Spider and Howling Monkeys of Santa Rosa National Park, Costa Rica," *Behaviour* 105, nos. 1–2 (1988): 99–116.

27. "*Bactris major*," Palmpedia, http://www.palmpedia.net/wiki/Bactris_major (accessed April 5, 2019).

28. "Red Fire Ants—Beware!" Caribbean Critters, https://ambergriscaye.com/critters /fireant.html (accessed April 5, 2019).

29. Jonathan A. Campbell and William Lamar, *The Venomous Reptiles of the Western Hemisphere* (Ithaca, NY: Comstock Publishing Associates/Cornell University Press, 2004).

30. Ibid.

31. "Venomous Snakes of Belize," Cayo Animal Welfare Society, http://www.cayo animalwelfaresociety.org/knowledge-base/venomous-snakes-of-belize/ (accessed April 5, 2019).

32. Ibid.

33. Peter Tyson, *The Eighth Continent: Life, Death and Discovery in a Lost World* (New York: William Morrow, 2000).

34. Ibid.

35. J. C. Masters, M. J. De Wit, and R. J. Asher, "Reconciling the Origins of Africa, India and Madagascar with Vertebrate Dispersal Scenarios," *Folia Primatologica* 77, no. 6 (2006): 399.

36. Tyson, *The Eighth Continent.*

37. Masters et al., "Reconciling the Origins of Africa, India and Madagascar."

38. Peter M. Kappeler, "Lemur Origins: Rafting by Groups of Hibernators?" *Folia Primatologica* 71, no. 6 (2000): 422–25.

39. George Gaylord Simpson, "Probabilities of Dispersal in Geologic Time," *Bulletin of the American Museum of Natural History* 99, no. 3 (1952): 227–34.

40. Ellen J. Censky, Karim Hodge, and Judy Dudley, "Over-Water Dispersal of Lizards Due to Hurricanes," *Nature* 395, no. 6702 (1998): 556.

41. J. Stankiewicz, et al., "Did Lemurs Have Sweepstake Tickets? An Exploration of Simpson's Model for the Colonization of Madagascar by Mammals," *Journal of Biogeography* 33, no. 2 (2006): 221–35.

42. Ibid.

43. Robert A. McCall, "Implications of Recent Geological Investigations of the Mozambique Channel for the Mammalian Colonization of Madagascar," *Proceedings of the Royal Society of London. Series B: Biological Sciences* 264, no. 1382 (1997): 663–65.

44. Ibid.

45. Peter M. Kappeler, "Lemur Origins: Rafting by Groups of Hibernators?" *Folia Primatologica* 71, no. 6 (2000): 422–25.

46. Mittermeier et al., *Lemurs of Madagascar*, 481.

47. Ibid.

48. Ibid.

49. N. Andriaholinirina, "Van der Decken's Sifaka," *The IUCN Red List of Threatened Species*, https://www.iucnredlist.org/species/18357/16116046 (accessed April 5, 2019).

50. Ibid.

Chapter 12

1. T. Leuteritz et al., "Madagascar Big-Headed Turtle, *Erymnochelys madagascariensis*," *The IUCN Red List of Threatened Species*, 2008: e.T8070A97396666. http://dx.doi.org/10.2305/IUCN.UK.2008.RLTS.T8070A12884059.en (accessed April 5, 2019).

2. "The World's Top 25 Most Endangered Turtles," Turtle Conservation Fund, 2003, http://www.turtleconservationfund.org/wp-content/uploads/2008/02/top25turtlesprofiles.pdf (accessed April 5, 2019).

3. Ibid.

4. Christoph Schwitzer et al., eds., *Lemurs of Madagascar: A Strategy for Their Conservation 2013–2016* (Bristol, UK: IUCN SSC Primate Specialist Group, Bristol Conservation and Science Foundation, and Conservation International, 2013).

5. Miguel Pedrono and Lora L. Smith, "Overview of the Natural History of Madagascar's Endemic Tortoises and Freshwater Turtles: Essential Components for Effective Conservation," in *Turtles on the Brink in Madagascar: Proceedings of Two Workshops on the Status, Conservation, and Biology of Malagasy Tortoises and Freshwater Turtles*, ed. Christina M. Castellano et al. (Lunenburg, MA: Chelonian Research Foundation, 2013), 59–66.

6. Ibid.

7. Dennis M. Hansen et al., "Ecological History and Latent Conservation Potential: Large and Giant Tortoises as a Model for Taxon Substitutions," *Ecography* 33, no. 2 (2010): 272–84.

8. Eric P. Palkovacs, "The Evolutionary Origin of Indian Ocean Tortoises (*Dipsochelys*)," *Molecular Phylogenetics and Evolution* 24, no. 2 (2002): 216–27.

9. Brooke E. Crowley, "A Refined Chronology of Prehistoric Madagascar and the Demise of the Megafauna," *Quaternary Science Reviews* 29, nos. 19–20 (2010): 2591–603.

10. David A. Burney et al., "A Chronology for Late Prehistoric Madagascar," *Journal of Human Evolution* 47, nos. 1–2 (2004): 25–63.

11. Russell A. Mittermeier et al., eds., *Lemurs of Madagascar*, 3rd edition (Arlington, VA: Conservation International, 2010).

12. N. Andriaholinirina et al., "Van der Decken's Sifaka, *Propithecus deckenii*," *The IUCN Red List of Threatened Species*, 2014, https://www.iucnredlist.org/species/18357/16116046 (accessed April 12, 2019).

13. "Madagascar: National Education Profile 2014 Update," Education Policy and Data Center, https://www.epdc.org/sites/default/files/documents/EPDC%20NEP_Madagascar.pdf (accessed April 5, 2019).

14. Helen Chapin Metz, ed., *Madagascar: A Country Study* (Washington, DC: GPO for the Library of Congress, 1994).

15. Ibid.

16. Ibid.

17. "The Children: Primary School Years," UNICEF Madagascar, https://www.unicef.org/madagascar/eng/children_6478.html (accessed April 5, 2019).

18. Jeanne Altmann, "Observational Study of Behavior: Sampling Methods," *Behaviour* 49, nos. 3–4 (1974): 227–66.

19. Ibid.

20. J. R. Napier and Colin Peter Groves, "Primate," *Encyclopedia Britannica*, January 16, 2019, https://www.britannica.com/animal/primate-mammal/Classification (accessed on April 5, 2019).

21. Frank Cuozzo and Nayuta Yamashita, "Impact of Ecology on the Teeth of Extant Lemurs: A Review of Dental Adaptations, Function, and Life History," in *Lemurs* (Boston, MA: Springer, 2006), 67–96.

22. Ibid.

23. Ibid.

24. Peter T. Ellison, *Reproductive Ecology and Human Evolution* (New York: Routledge, 2017), 478.

25. Ibid.

Chapter 13

1. David Werner, Carol Thuman, Jane Maxwell, *Where There Is No Doctor: A Village Health Care Handbook* (London: Macmillan Education, 1993).

2. Ibid., 26.

3. Ibid., 186.

4. William C. Shiel Jr., "Medical Definition of Malaria, falciparum," https://www.medicinenet.com/script/main/art.asp?articlekey=4256 (accessed April 5, 2019).

5. Richard Idro, "Cerebral Malaria: Mechanisms of Brain Injury and Strategies for Improved Neurocognitive Outcome," *Pediatric Research* 68, no. 4 (2010): 267–74.

6. Werner, Thuman, and Maxwell, *Where There Is No Doctor.*

7. "Madagascar: Overview," The World Bank in Madagascar, 2017, https://www.worldbank.org/en/country/madagascar/overview (accessed April 1, 2019).

Chapter 14

1. Russell A. Mittermeier et al., eds., *Lemurs of Madagascar*, 3rd edition (Arlington, VA: Conservation International, 2010), 636–38.

2. Ibid., 638.

3. "Sickle-Billed Vanga," Oiseaux-Birds, http://www.oiseaux-birds.com/card-sickle-billed-vanga.html (accessed on April 5, 2019).

4. Ian Sinclair, *Birds of the Indian Ocean Islands* (Cape Town, South Africa: Struik Nature, 2013).

5. John R. Platt, "Climate Change Could Wipe Out Amazing Baobab Trees in Madagascar," *Scientific American,* July 19, 2013, https://blogs.scientificamerican.com/ extinction-countdown/climate-change-baobab-madagascar/ (accessed April 5, 2019).

6. Carmen Liberatore, "Ankarafantsika National Park and the Legend of Lac Ravelobe," *Lemur Conservation Network*, August 18, 2015, https://www.lemurconserva tionnetwork.org/ankarafantsika-national-park-and-the-legend-of-lac-ravelobe/ (accessed April 6, 2019).

7. Ibid.

Chapter 15

1. Kasturi Haldar and Narla Mohandas, "Malaria, Erythrocytic Infection, and Anemia," *Hematology, American Society of Hematology, Education Program* (2009): 87–93.

Index

Note: Page references for figures are italicized.